The Botany
and Chemistry
of Hallucinogens

Publication Number 843

AMERICAN LECTURE SERIES®

A Monograph in

The BANNERSTONE DIVISION *of*
AMERICAN LECTURES IN LIVING CHEMISTRY

Edited by

I. NEWTON KUGELMASS, M.D., Ph.D., Sc.D.

*Consultant to the Departments of Health and Hospitals
New York City*

The Botany and Chemistry of Hallucinogens

By

RICHARD EVANS SCHULTES, Ph.D., M.H. (Hon.)

Professor of Biology
Director and Curator of Economic Botany
Botanical Museum of Harvard University
Cambridge, Massachusetts

and

ALBERT HOFMANN, Ph.D., Dr. Pharm. H.C., Dr. Sc. Nat. H.C.

Head of the Pharmaceutical-Chemical Research Laboratories
Division of Natural Products, Sandoz Ltd.
Basel, Switzerland

With a Foreword by

Heinrich Klüver, M.D. (Hon.), Ph.D.

Professor Emeritus
Division of the Biological Sciences
University of Chicago
Chicago, Illinois

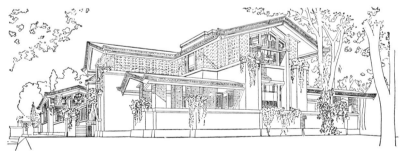

CHARLES C THOMAS · PUBLISHER
Springfield · Illinois · U.S.A.

Published and Distributed Throughout the World by

CHARLES C THOMAS · PUBLISHER

BANNERSTONE HOUSE

301–327 East Lawrence Avenue, Springfield, Illinois, U.S.A.

© *1973, by* CHARLES C THOMAS · PUBLISHER

ISBN 0–398–02401–4

Library of Congress Catalog Card Number: 72–187675

With THOMAS BOOKS *careful attention is given to all details of
manufacturing and design. It is the Publisher's desire to present books
that are satisfactory as to their physical qualities and artistic possibilities
and appropriate for their particular use.* THOMAS BOOKS *will be true
to those laws of quality that assure a good name and good will.*

Printed in the United States of America

K-8

EDITOR'S FOREWORD

Our Living Chemistry Series was conceived by Editor and Publisher to advance the newer knowledge of chemical medicine in the cause of clinical practice. The interdependence of chemistry and medicine is so great that physicians are turning to chemistry, and chemists to medicine, in order to understand the underlying basis of life processes in health and disease. Once chemical truths, proofs and convictions become sound foundations for clinical phenomena, key hybrid investigators clarify the bewildering panorama of biochemical progress for application in everyday practice, stimulation of experimental research, and extension of postgraduate instruction. Each of our monographs thus unravels the chemical mechanisms and clinical management of many diseases that have remained relatively static in the minds of medical men for three thousand years. Our new Series is charged with the *nisus élan* of chemical wisdom, supreme in choice of international authors, optimal in standards of chemical scholarship, provocative in imagination for experimental research, comprehensive in discussions of scientific medicine, and authoritative in chemical perspective of human disorders.

Dr. Schultes of Cambridge, Massachusetts, and Dr. Hofmann of Basel, Switzerland, combine their botanical and chemical knowledge to integrate interdisciplinary knowledge and international wisdom about powerful alkaloids from primitive cultures to modern times. There are only 60 species of hallucinogens out of about 600,000 plant species with an unexplained concentration of these alkaloids in the New World, with hallucinogenic plants in both hemispheres never used as narcotics. And so the authors conduct the reader through the classic edifice of clinical alkaloids into world areas where probing is now in progress, to temper the customary severity of the sciences insofar as is compatible with clarity of thought. They know how to put what they have to say as if it had never been said before. It is not a textbook, a table d'hôte to which one may sit down and satisfy his hunger for drug information with no thought of the agricultural processes which gave rise to the raw materials, nor of the laboratory procedures which converted them into drugs. It is an authoritative work designed as an introduction to research and as a guide to doctors to use hallucinogens effectively. Once chemical understanding is clear, clinical application is usually easy.

Naturally occurring hallucinogens have been used in religious practices for centuries and in psychiatric practice since the beginning of the century; artificial hallucinogens have come into use during this generation. They produce extraordinary disturbances of perception—disordered sense of time, visual hallucinations, mystical experiences and even transcendental sensations—all disrupting the function of the ego. Unlike cocaine derivatives, none is addicting; unlike barbiturates, none depresses brain functions. Hallucinogenic drugs alter human consciousness in man's groping toward memory control, reshaping behavior patterns through reinforcement schedules paralleling man's vertical extension into outer and inner space. The results of this inward exploration may be infinitely more powerful than voyages to planets or to the bottom of the Challenger Deep. The territory is virtually unexplored. But do the construction potentials of hallucinogens outweigh their admitted hazards? There are really no safe hallucinogens; there are only safe physicians. Even in treating patients, doctors will have to be extremely cautious not to transgress the rule of *primum non nocere*.

> *But words are things, and a small drip of ink,*
> *Falling like dew upon a thought, produces*
> *That which makes thousands, perhaps millions, think.*

I. Newton Kugelmass, M.D., Ph.D., Sc.D., *Editor*

FOREWORD

I FEEL greatly honoured by the request of the authors to furnish a few preliminary remarks of introduction to this book. At the same time, I should confess, I am somewhat reluctant to undertake such a task since the subject of "hallucinogens" is related to one of the most complex and difficult subjects—to the world of hallucinations and other subjective phenomena. In trying to cope with this "hallucinatory" world, previous investigators have left us with categories and terms such as *positive* and *negative hallucinations, pseudohallucinations, déjà vu phenomena, derealization phenomena, illusions, visualizations, dreams, dreamy states, eidetic images, eidetic images with reality character, hypnagogic images, memory-images, projected memory-images, memory-after-images, pseudo-memory-images, "phantastic visual phenomena," re-perceptions,* and *Sinnengedächtnis,* to name only a few of the "technical" terms used in these researches. Raoul Mourgue, after analyzing some 7,000 publications on the subject of hallucinations, was forced to conclude in 1932, in his *Neurobiologie de l'hallucination,* that all the facts and observations then available could not furnish an adequate basis for a *theory* of hallucinations. Fortunately, the reader of the present volume, *The Botany and Chemistry of Hallucinogens,* will readily see that it is not the aim of the authors to deal with and attempt a resolution of such psychological intricacies; their aim is obviously a more important one: to start at the "beginnings" by furnishing a detailed presentation of such botanical and chemical facts and considerations as are essential for a scientific study of hallucinogens. Certainly, such a point of departure will ultimately be essential also for an understanding of the mechanisms of hallucinations.

vii

No one will doubt that the two authors are particularly well qualified to write this particular book. Only experts in both botanical and chemical matters could deal with all the ramifications and complexities of the problems presented here. Personally, I merely claim a long-continuing and deep interest in the various problems so competently discussed in this volume. This interest of mine has been greatly strengthened by my belief that the authors of this book are rather unique in the annals of science: the one, Richard Evans Schultes, for remaining in the jungle for some 12 years to acquire the tools of his trade and his facts; the other, Albert Hofmann, for constantly following the trail he discovered on April 16, 1943, a day that was destined to launch, not one thousand but several, perhaps even ten thousand, publications on LSD.

At a time when drugs and drug-produced experiences have become a national and international problem, the reader may easily forget that not so long ago, during the first decades of this century, only a few investigators in various countries were interested in a scientific study of hallucinogens. As it happened, on an October day in 1925, I introduced myself to the world of hallucinogens by consuming some "mescal buttons" in one of the laboratory buildings of the University of Minnesota, not for the sake of "consciousness expansion" or other unique experiences but to test a new tool possibly useful in studying various problems of the psychology and pathology of perception. For some years, I had been engaged in the study of certain types of pseudohallucinations (now generally referred to as *eidetic phenomena* or *eidetic imagery*), studies which had taken me into different geographic districts in California, New York, West Virginia, Ohio, and Louisiana. It was a remark found by chance in the literature that *Lophophora williamsii* would provide the possibility of producing eidetic phenomena in noneidetic individuals (I happen to be such an individual) that led me to the use of peyote in my 1925 experiment and to the results which I subsequently reported in the literature (*Am. J. Psychol.*, 37: 502–515, 1926). It also led, somewhat unexpectedly, to a stream of visitors ranging from organic chemists to psychiatrists and anthropologists who appeared in my laboratory or home; visitors who for one reason or other had become interested in the world of mescaline-produced phenomena.

Thus, I recall with pleasure the long and profitable discussions with researchers such as Gordon Alles, Paul Radin, J. S. Slotkin, W. Mayer-Gross, A. Hoffer, and H. Osmond. I also recall that in one of these discussions, Paul Radin, who had made a special study of the peyote cult, offered to send "Crashing Thunder" for some psychological testing to my laboratory (Crashing Thunder being the Winnebago Indian whose autobiography he had edited). I declined his kind offer for various "psychological" reasons, believing that my Culver Hall laboratory at the University of Chicago would not provide the proper background for psychologically examining a Winnebago Indian. . . .

Some 50 years ago there seemed to be relatively few investigators who were concerned with scientific approaches to the study of hallucinogens or who were using drugs in exploring mechanisms of normal and abnormal behavior. Consequently, I could optimistically assume that it would be relatively easy to contact at least the most important of these researchers and visit their laboratories and institutions. In pursuit of such an idea, I made, for instance, a "pilgrimage" to the Maison Nationale de Charenton, an institution associated with such famous names as Esquirol and Moreau de Tours. My host was Henri Baruk who, in his drug researches, happened to be a collaborator of H. de Jong in Holland (whom I subsequently visited in his laboratory in Amsterdam). Incidentally, it was Henri Baruk who, in 1958, founded the Société Moreau-de-Tours and who, in 1962, played a leading role in publishing the Annales Moreau de Tours. It was on the same European trip that I had the pleasure of spending a day in Freiburg i.B. with Kurt Beringer, whose book, *Der Meskalinrausch*, in the opinion of Bo Holmstedt, is to mescaline what Moreau's book, *Du Hachisch et de l'aliénation mentale*, is to hasheesh.

As to American investigators in the 20's, I like to recall that, during my years at Columbia University (1926–1928), I managed to discover the whereabouts of William J. A. M. Maloney who, together with A. Knauer, had studied the "psychic action" of mescaline with special reference to "the mechanisms of visual hallucinations" in Kraepelin's clinic in Munich. This was no doubt the most thorough experimental study of mescaline effects carried out before World War I, and it was carried out with the methods of experimental psychology, methods with which Kraepelin had become familiar in the Leipzig laboratory of Wundt. It is probably not generally known that Kraepelin, who became famous for initiating a new era in psychiatry, thought most highly of his contributions to experimental psychology; his "deepest love" belonged to this field. When I discovered that Maloney was practicing clinical neurology in New York City, I called on him but unfortunately found his waiting room full of patients. However, when he heard from the nurse that I wanted to talk about mescaline, he took me immediately into his office and talked for at least an hour about "the happiest year" in his life, namely, the year he and Knauer had spent in Kraepelin's clinic studying the psychological effects of mescaline poisoning. He also explained to me the sad reason that a detailed report of these remarkable pre-World War I experiments, aside from a preliminary report in 1913, would never be published (and to my knowledge has never been published).

It was at the time of my visit to Maloney that I found another New York investigator who had also been concerned with hallucinogens and had, in fact, written a handbook article on mescal buttons. This was Henry Hurd Rusby (1855–1940) who was associated with the Department of Pharmacy at Columbia University. His interests were wide ranging—from the morphology and histology of plants to the properties and uses of drugs; his work had brought him many medals and honors. When I called on him one late afternoon, I found him in his office surrounded by half a dozen black dogs. In the course of the conversation he soon warned me (and any other scientist) against studying unknown drugs by experimenting on oneself. Finally, he took me into the next room filled with large boxes containing materials which he had brought home from various expeditions. He would probably never unpack them, he said, or work up their contents; in fact, he was now raising dogs and asked me to join him in giving his dogs an airing on Broadway (which I did).

When, many years later, I did some reminiscing about my

Page remarked that this discovery would certainly have "fascinated investigators and clinicians alike" if they had known about it at that time. At the present time, the historian will have to record that interest in psychochemistry has spread to fields far removed from medicine and biology. The scientific activities carried on by a man like Max Knoll during the last decade of his scientific career strikingly exemplify and illustrate such developments. A psychologist or psychiatrist may even argue that the exploration of the "psycho" in psychochemistry or psychopharmacology has not kept pace with advances along chemical and pharmacological lines, not to mention the progress in electronics or physics. Furthermore, it is apparent that the well-nigh frenetic research activities in the field of psychoactive drugs have frequently been pursued without considering recent advances along ethnopharmacological and ethnobotanical lines.

Fortunately, researchers of the future ready to explore further the world of psychotomimetics and hallucinogens will now have a chance of avoiding the sad discovery at the end that they have ignored relevant facts or concepts and having to cry peccavi for such a distressing or even fateful oversight. They may easily avoid, it might be suggested, such a course of events by studying *The Botany and Chemistry of Hallucinogens* by Richard Evans Schultes and Albert Hofmann before embarking on their researches! This is, I believe, sound advice.

HEINRICH KLÜVER

PREFACE

A NYONE might justifiably ask, "Why another book on the hallucinogens?" The past decade has witnessed a steady proces-
sion of volumes in the many disciplines touched upon by these
most remarkable narcotics, not to mention the thousands of articles in learned journals. Some of these works have been good,
some mediocre, some poor; there have been specific and there
have been comprehensive treatments; a few have been impartial
scientific treatises, many have been emotional "potboilers." The
1960's will most certainly be remembered as a period in which
hallucinogens have called forth in both technical and popular
literature a plethora of productions. Why, then, still another
book?

We believe that there is great need—nay, even a vital urgency
—for a simple treatment that starts the study of hallucinogens at
the very basics. A survey of the vast literature indicates a conspicuous dearth of books which start at the beginning of any
serious effort to understand hallucinogens and their impact on
human affairs—in primitive cultures of hinterland regions or in
sophisticated societies of the western world.

Most hallucinogens are of vegetal origin. Consequently, the
first step in any consideration of hallucinogens must be botanical.
Even though specialists know it, they sometimes tend not to
realize that hallucinogens are by and large plant products (that
means natural products) and have been available since long before the mind of man was capable of recognizing their utility.
Consequently, a sound understanding of their value and effectiveness, their dangers or innocuousness, lies in an evaluation of their
botanical identity and, concomitantly, a clear and thorough

knowledge of their chemical composition. These two phases of technical appreciation of hallucinogens must be clarified before any studies in such fields as psychopharmacology, physiology and the behavioural sciences can be expected to yield significant results.

Consequently, we—a botanist and a chemist, both of whom have worked with hallucinogens for long periods—have sought to present a brief, straightforward book oriented primarily towards the botany and chemistry of hallucinogens. It most certainly will fall far short of an ideal, inclusive text on these two basic fields of hallucinogenic research. We trust, however, that here specialists not only in botany and phytochemistry, but also in the many other disciplines that impinge upon the growing study of the hallucinogens, may find basic information simply set forth, free from the encumbrances of extraneous discussion and all-embracing discourse, and may use this information for the furtherance of their own scientific endeavours.

RICHARD EVANS SCHULTES
ALBERT HOFMANN

INTRODUCTION

W<small>E HAVE TRIED</small> to keep the plan of our book simple. Oftentimes the information available does not permit us to assert positively that a toxic or narcotic plant employed in a primitive society is used because of hallucinatory effects. In other instances, the data at hand seem to indicate that the use is for the purpose of inducing hallucinations, yet we know of no psychotomimetic constituent in the plant. In a few cases, both the utilization of a plant for psychoactive effects and its chemical composition are equally doubtful or uncertain.

Believing that the purposes of future investigations are better served by including rather than excluding these borderline examples, we have set them apart in a section of the book separate from the well known, adequately understood hallucinogenic plants, the use of whch is based definitely upon the search for visual and/or other types of hallucinations.

In the section of this book dealing with plants of doubtful hallucinogenic properties, no chemical formulae are given. The constituents of such plants are mentioned only by their names and reference to their chemistry is made in bibliography. Chemical compounds with nonspecific hallucinogenic activity (e.g. ibogaine) of which the hallucinogenic property is only a side effect of other main pharmacological activities are characterized by their structural formulae, but without description of their synthesis, which is referred to only in the bibliography. Detailed chemistry will be reported only for the specific hallucinogens.

Throughout the book, we have employed the term *hallucinogen* or *psychotomimetic* to mean either the active chemical principles or the plant containing them, or crude extracts of such plants.

Our reason for not reserving these terms merely for chemical compounds is simply that, in primitive societies especially, where these psychoactive agents find their primary use, rarely if ever are the responsible chemical constituents isolated from the plant and taken as such. The normal procedure amongst aborigines is the employment of the crude plant in its natural state or, at best, as a crude decoction, infusion or powder of the vegetal material. It is only rarely and in sophisticated western cultures that purified chemical compounds are isolated and taken for hallucinogenic purposes.

In order to provide the reader with an understanding of the botanical interrelationship, we have, in discussing the major hallucinogenic plants, felt constrained to present very briefly basic information on each family and genus considered, and a rather detailed description of the species involved. Sometimes—as with the sacred hallucinogenic mushrooms of Mexico and the myristicaceous snuffs of the Amazon—a number of closely related species are employed, and in these cases we have offered a description of only one species, choosing what appears to be the most important species.

One of the contributions of our book which, we believe, many colleagues may find most useful is the bibliography. Consequently, we have aimed towards a rather comprehensive and inclusive list of references. It remains, however, perfectly obvious that a complete bibliography of such a fast moving field is not possible. We have, therefore, had occasionally to exercise our judgment in what to include and what to exclude in accordance with the aim of our book.

R.E.S.

A.H.

ACKNOWLEDGMENTS

THERE ARE so many colleagues and friends who have contributed to the elaboration of this book, both in the botanical and chemical aspects of it, that we find it difficult to thank adequately every one of our helpers.

To those colleagues who have themselves worked on hallucinogenic plants or chemical constituents we owe our appreciation for many suggestions and favours: Professors Roger Heim (Muséum d'Histoire Naturelle, Paris), Bo Holmstedt (Karolinska Institutet, Stockholm), Robert F. Raffauf (Northeastern University, Boston, Mass.), Heinrich Klüver and Norman Farnsworth (University of Chicago, Chicago, Ill.), Peter Waser (University of Zurich, Zurich, Switzerland), Weston La Barre (Duke University, Durham, N. C.), Doctors Tony Swain and Patrick Brenan (Royal Botanic Gardens, Kew), György-Miklos Ola'h (University of Laval, Quebec, Canada), Nathan S. Kline (Rockland State Hospital, Orangeburg, N. J.), Humphrey Osmond (New Jersey Neuropsychiatric Institute, Princeton, N. J.), Stig Agurell (University of Uppsala, Stockholm), Jan-Erik Lindgren (Karolinska Institutet, Stockholm), Siri von Reis Altschul (Harvard University, Cambridge, Mass.), Mr. S. Henry Wassén (Etnografisk Museet, Gothenburg) and R. Gordon Wasson (Harvard University, Cambridge, Mass.). Prof. Peter Furst (University of California at Los Angeles, Cal.) and Dr. Donovan S. Correll (National Science Foundation, Washington, D. C.) have been extremely helpful with suggestions and willingness to procure photographs and plant materials.

The staffs of several libraries have been most generous with their time, but we must thank especially Miss Esther Reynolds and Mrs. Lillian Hanscom of the Library of Economic Botany at the Harvard Botanical Museum for many favours.

Several artists have assisted materially in providing illustrative material of the hallucinogenic plants, especially Mr. Gordon W. Dillon, Mr. Elmer W. Smith, Mr. Joshua Clark, Mrs. Irene Brady Kistler, and Miss Judith Gronim who have worked at the Botanical Museum of Harvard University.

Professors Conrad H. Eugster (University of Zurich), J. W. Fernandez (Dartmonth College) A. Danin (Hebrew University of Jerusalem), Dr. José Cuatrecasas (Smithsonian Institution; Washington D. C.), Mr Homer V. Pinkley and Mr. Timothy Plowman (Harvard University), Mrs. Julia F. Morton (University of Miami), Professor James W. Walker (University of Massachusetts) and Dr. Melvin L. Bristol have generously permitted us to publish photographs and illustrations which they prepared during fieldwork and studies. Dr. George Lawrence (Hunt Botanical Library, Pittsburgh, Penn.), Professors Roger Heim, Bo Holmstedt, and György-Miklos Ola'h have made available portraits of value to the historical portion of this book.

Much of the information presented in this book has been gathered and brought together with the aid of grant number LM-GM 00071–01 from the National Institute of Health.

Finally, our debt of gratitude to Mrs. Helen de Huarte (Beltsville, Md.), who meticulously typed the manuscript, can hardly be exaggerated.

R.E.S.
A.H.

CONTENTS

The Botany
and Chemistry
of Hallucinogens

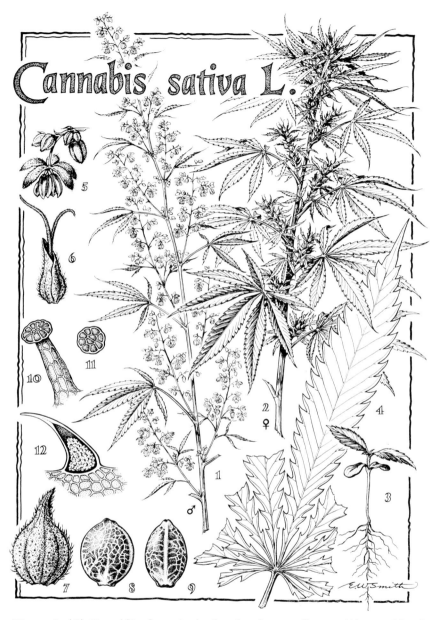

Cannabis sativa L.

Figure 1. (1) Top (distal portion) of male plant, in flower, (2) top (distal portion) of female plant, in fruit, (3) seedling, (4) leaflet from a large, 11-parted leaf, to show leaf variability, (5) portion of a staminate inflorescence, with buds and a mature male flower, (6) female (carpellate) flower, with stigmas protruding from enveloping hairy bract, (7) fruit enclosed in persistent hairy floral bract, (8) fruit, lateral view, (9) fruit, end view (sagittal plane), (10) glandular hair, with multicellular stalk, (11) glandular hair with short, one-celled, invisible stalk (sometimes called a sessile glandular hair), (12) rigid non-glandular hair containing a cystolith of calcium carbonate. Drawn by E. W. Smith.

Chapter I

HALLUCINOGENIC OR PSYCHOTOMIMETIC AGENTS: WHAT ARE THEY?

Agents that cause visual, auditory, tactile, taste and other hallucinations or that induce artificial psychoses have been known undoubtedly since earliest human experimentation with the vegetal environment.

Their use goes so far back into prehistory that it has been suggested that even the idea of the deity may have arisen as a result of their physiological effects.[426] Although man in all primitive cultures tried to find direct palliatives or cures for his ills, the psychic effects of drugs were often far more important to him than the purely physical. It is easy to understand, since in most, if not all, primitive cultures, sickness and death itself are attributed usually to supernatural forces entering the body. Witchcraft, aided by communion with spirit forces, was the principal tool in diagnosing and treating disease.[349]

What easier way to contact the spirit world than the use of plants with strange, unearthly, psychic effects capable of freeing man from the prosaic confines of this mundane environment and carrying him temporarily to fascinating worlds of indescribably ethereal wonder? Primitive man sought after these plants and put an extraordinary value on them. Narcotics, especially those now called hallucinogens, were his medicine *par excellence* and became fast fixtures of his magic and religion, the bases of his medical practices.[371]

How could this come to pass? Man saw, experimented with and came to know the thousands of different plants in his surroundings. He tried ingesting them all. Most were innocuous; a few pleased his taste; some nourished him; a goodly number of them made him ill; sundry ones relieved pain and suffering; a few killed him outright; but a very few had weird and unearthly effects on his body and mind. His only plausible explanation of such strangely unreal, psychic powers ascribed to these species a resident divinity or spirit. The plants were exalted to a highly sacred place and usually were reserved for a sacred role in magico-religious rites.[208,346]

Modern man now knows that these "resident divinities" are chemical substances. During the last eighty years—but more particularly during the last two decades—pharmaceutical research has uncovered an astonishing array of chemical compounds, many from plant sources, capable of extraordinarily affecting the psychic functions. The discovery of these new and potent psychotropic compounds created a new field of medical science that has come to be known as *psychopharmacology*.

Several books may be cited as having laid the bases for modern psychopharmacology. In 1855, Ernst Freiherr von Bibra published the first book of its kind, *Die narkotischen Genussmittel und der Mensch*,[38] in which he considered some 17 plant narcotics and stimulants, urging chemists to study assiduously a field so promising for research and so fraught with enigmas.

Mention should be made of the fundamental investigations of Emil Kraepelin, who, in his monograph *Ueber die Beeinflussung einfacher psychischer Vorgänge durch einige Arzneimittel*,[203] published in 1892, delineated much of what later led to psychopharmacology.

Half a century after von Bibra's work and undoubtedly sparked by it, another outstanding book appeared in 1911 which was C. Hartwich's *Die menschlichen Genussmittel*.[133] This extensive volume considered at great length and with interdisciplinary emphasis about 30 vegetal narcotics and stimulants and mentioned others in passing. Pointing out that von Bibra's pioneer work was out of date, that botanical and chemical research on these curiously active plants had scarcely begun in 1855, he optimistically maintained that, by 1911, such studies were either well under way or had already been completed.

Thirteen years later, in 1924, perhaps the most influential figure in psychopharmacology, Louis Lewin, published a book of extraordinary interdisciplinary depth: *Phantastica—die betäubenden und erregenden Genussmittel.*[218] It was shortly translated into several languages, the earliest English edition (*Phantastica: Narcotic and Stimulating Drugs—Their Use and Abuse*) appearing in 1931, with a new edition coming out in 1964. It presented the total story of some 28 plants and a few synthetic compounds which are employed around the world for their stimulating or inebriating properties, emphasizing their importance to research

Figure 2. Carl Hartwich.

in botany, ethnobotany, chemistry, pharmacology, medicine, psychology, psychiatry, as well as to ethnology, history and sociology. Lewin humbly wrote that "the contents of this book will provide a starting point from which original research in the above-mentioned departments of science may be pursued."[218]

In the years between the works of Hartwich and Lewin, an American ethnobotanist, William E. Safford, began to focus scientific attention on the wealth of narcotic plants employed in primitive societies of the New World. Calling attention to numerous enigmas in the botanical identification of some of the hallucinogens of ancient American cultures, he must be credited with pioneering research into the rich field of narcotics in the Western

Figure 3. Louis Lewin.

Hemisphere, even though several of his own attempts at identifying them later were shown to be erroneous.[301-307]

From the point of view of physiology and psychology, perhaps one of the most influential books, and one that has indeed been a milestone in psychopharmacology, is Kurt Beringer's *Der Meskalinrausch*,[36] published in 1927.

From about 1930 on, interdisciplinary activity in psychopharmacology began uninterruptedly to increase. Many clarifications and amplifications of older knowledge were made, and new discoveries in sundry fields followed one another in close succession. The hallucinogenic or psychotomimetic plants and plant constituents have, especially in the past decade, attracted attention far beyond the several disciplines of science.

Figure 4. William E. Safford.

It is in many ways difficult, often impossible, to delimit accurately what we mean by *hallucinogens*, primarily because their biological activity is so diverse and complex.

Lewin grouped psychoactive plants in five categories: *excitantia, inebriantia, hypnotica, euphorica, phantastica*. None has stirred deeper interest than the phantastica—the drugs that produce hallucinations. Unfortunately, this term has not been generally adopted; in fact, in English-speaking countries, it is unknown.

As in every fast developing field of study, a burgeoning nomenclature has grown up around these hallucinatory agents. They have variously been called *delirients, dysleptics, mysticomimetics, phantasticants, psychotica, psychoticants, psychogens, psychotogens, psychodysleptics, psychotaraxics, eidetics, schizogens, hallucinogens, psychosomimetics,* and *psychotomimetics.* Most recently, the term *psychedelics* has found wide acceptance in the United States. It was meant to signify "mind manifesting" and was intended to convey the fact that the pattern of effects of these compounds is determined not so much by the pathological components as by a general activation and manifestation of psychic phenomena, which in no way may be considered entirely morbid and negative. However, the term *psychedelic* has now acquired secondary and even tertiary meanings in certain sections of modern society and, for this reason if for no other, has no value in scientific language.

None of these terms, to be sure, is wholly and always satisfactory. Even Lewin, when he coined the term *phantastica,* was not completely satisfied with it, complaining that it "does not cover all that I should wish it to convey." [218] The same might be said of any of the words that have been applied to these curious drugs.

Lewin defined his group *phantastica,* from the point of view of a toxicologist, as "hallucinating substances . . . a number of substances of vegetable origin, varying greatly in their chemical constitution, and to these belongs in its proper sense the name Phantastica or Drugs of Illusion. The representatives of this group . . . bring about evident cerebral excitation in the form of hallucinations, illusions and visions. These phenomena may be accompanied or followed by unconsciousness or other symptoms of altered cerebral functioning." [218]

Two terms are easily understood and now widely employed, albeit still, in some respects, deficient. Since we believe that they are the best that are available, we intend not to wallow in sterile semantics, but to use *hallucinogens* ("giving rise to hallucinations") and, to a lesser extent, *psychotomimetics* ("mimicking a psychosis") in this book. The term *hallucinogens* emphasizes perceptual changes, whereas alterations of thought or mood may sometimes be much more important and significant; hallucinations, while most commonly visual, may involve any of the senses, leading to auditory, tactile and olfactory aberrations. The term *psychotomimetics* emphasizes the often near-psychotic or pseudopsychotic state induced by these agents, but a mimicking by general activation and manifestation of psychic phenomena is in no way entirely morbid and negative. The exceptional psychic states induced by psychotomimetics only partially correspond to the pathological pattern of schizophrenia, and these agents may not necessarily make the partaker psychotic.

A recent medically oriented discussion of chemical psychoses considered the definition of psychotomimetic drugs as follows:

> While any definition of the term *psychotomimetic drugs* is bound to be arbitrary, one can limit the field somewhat if the following criteria are used: (a) In proportion to other effects, changes in thought, perception and mood should predominate; (b) intellectual or memory impairment should be minimal with doses producing the above mental effects; with large doses these may occur; (c) stupor, narcosis or excessive stimulation should not be an integral part of the action; (d) autonomic nervous system side-effects should be neither disabling nor severely disconcerting; (e) addictive craving should be minimal. Even with criteria such as these, drugs admissible to the list may vary, depending upon the investigator. Indeed, one can scarcely get any agreement upon the term used to describe this class of drugs, as many objections to the most likely used term, *psychotomimetics,* can be offered.[177]

Another recent monograph, more psychiatrically oriented, acknowledges that the term *hallucinogen* is not wholly satisfactory because ". . . it overemphasizes the perceptual elements of the response to these drugs, and perceptual changes are often minor; changes in thought and mood are much more important . . . Hallucinogens are . . . chemicals which, in nontoxic doses, produce changes in perception, in thought, and in mood, but which seldom produce mental confusion, memory loss, or disorientation for person, place, and time."[154]

Hallucinogens or psychotomimetics, in effect, are agents which, in nontoxic doses, produce, together or alone, changes in perception, thought and mood, without causing major disturbances of the autonomic nervous system. A variety of hallucinations may be characteristic, especially with high doses. Disorientation, loss or disturbance of memory, excessive impairment of intellectual powers, hyperexcitation or stupour or even narcosis may be experienced only under excessive doses and cannot, therefore, be considered characteristic. Addiction is unknown with these drugs.

The foregoing definition, as with most definitions, may be arbitrary in some aspect, and is not completely inclusive or exclusive; exceptions to one or more of the conditions enumerated may not be difficult to find. It does, however, cover the characteristic activity of hallucinogens or psychotomimetics and certainly does exclude narcotics such as morphine and cocaine and their derivatives, as well as anaesthetics, analgesics and hypnotics. The definition, in short, clearly delineates Lewin's category of phantastica, with which hallucinogen and psychotomimetic must be considered synonymous.

The psychic changes and the attendant abnormal states of consciousness induced by hallucinogens differ so utterly from ordinary experiences of the outer and inner world that they cannot be described in the usual language of the daily pattern of the outer and inner universe. The profound changes in conception of the universe towards either the diabolical sphere or celestial transfiguration may be explained by alterations in space and time perception—the two basic elements of human existence. The experience of corporeity and the spiritual being may likewise be deeply affected. The partaker of an hallucinogen forsakes the familiar world and, yet in full consciousness, embraces a kind of quasi-dream world operating under other standards, strange dimensions and in a different time.

Time may frequently seem to stand still, not to exist. The familiar, daily environment appears in a new and often glorious light. Forms and colours are changed or acquire a new, sometimes far-off significance. Ordinary objects lose any symbolic character, are detached, radiate their own intense entity. Colours usually become richer, transparent, radiating from the inside. Visual and auditory hypersensitivity are common and often, especially with high doses of the psychotomimetic agent, lead to hallucinations.

Hallucinogens, whether taken by members of a primitive society or by groups in sophisticated cultures, are used as a means of escaping from reality, as we commonly understand reality. Whether it be the witch doctor who seeks communion with the spirit world or the artist hoping for new horizons—both act basically to leave the confines of daily mundane living. Although many peoples have known and used hallucinogenic plants from earliest stages in their cultural development, others have sought release from reality through slow and painful avenues such as starvation and self-torture. Be it the chemical method of a South American witch doctor or the fasting of an early Christian mystic, the end results may differ little: both individuals experience an exceptional psychic state. Perhaps superficially at least the only difference that can be noted lies in the relatively greater ease of the chemical avenue (the way of the hallucinogens or psychotomimetics), which may lead to and act through changes in the biochemistry of the body. It is still controversial whether or not drug-induced experiences can be basically identical with the metaphysical insight claimed by some mystics—or merely are a counterfeit of them.

During the last 80 years, particularly within the past two decades, pharmaceutical and phytochemical research has succeeded in isolating from many of the so-called "magic plants" the active principles in chemically pure form. Their structures have been elucidated, and the important hallucinogenic agents have been synthesized. These important psychoactive compounds have gained a place of significance in psychopharmacological research, in experimental psychiatry, in psychoanalysis and even in psychotherapy.

In the future, psychotomimetic substances may serve to produce experimentally the kinds of psychoses—the so-called "model psychoses"—which will enable science to study biochemical or electrophysiological processes that might be allied to mental disorders. In psychoanalysis or psychotherapy, they may be utilized as potential therapeutic tools. The effects which would seem to give the psychotomimetics this potentiality may be summarized as follows. First, hallucinogens alter the patient's normal concept of the world, freeing him from his autistic fixation and his feeling of isolation. As a consequence of this psychic change, the patient may experience a better rapport with his physician. Second, psychotomimetics call up forgotten or repressed subconscious material. Through hallucinogenic agents, experiences, even those going back to early childhood which might often be the cause of mental disorders, may occasionally enter vividly into his consciousness; this recognition of the cause of the disorder is basic to any successful psychotherapeutic treatment. Discovery of conflicts that have led to mental illness may thus be speeded up, and the psychoanalytical treatment may possibly be shortened.

Without supervision, patients should not experiment with substances possessing such deep and fundamental effects as those that the hallucinogens have on the mind and body, for many of the effects, their intensity, duration and after effects, can in no way be foreseen. From a therapeutic point of view, such free use of hallucinogenic agents is not only useless but even dangerous. Hallucinogens simply are not of themselves medicines; they are aids or tools of therapy.

Chapter II

THE BOTANICAL DISTRIBUTION
OF HALLUCINOGENS

ALMOST ALL hallucinogens are of vegetal origin. The most potent psychotomimetic agent, LSD or lysergic acid diethylamide, although it is a synthetic compound not known to occur in plant tissue, is closely related to some of the natural hallucinogens of the Plant Kingdom, since the major part of its molecule (i.e. the lysergic acid radical) is of natural origin. Furthermore, the unsubstituted lysergic acid amide and other lysergic acid derivatives very closely related to LSD have been found in plants employed as hallucinogens.

How many species of plants are employed for their hallucinogenic properties throughout the world is not known with any certainty. How many species possess hallucinogenic principles but have never been used for their narcotic properties is known with even less certainty. Recent phytochemical investigations tend to suggest that psychotomimetic constituents are rather widely scattered throughout the Plant Kingdom in many species which man has either not discovered as hallucinogenic or which, for one reason or other, he has ignored and failed to utilize for purposes of intoxication.[339,340,347]

In the course of his one million years of existence, man must have experimented with a large percentage of plants in the environment. Botanists cannot definitely state how many plants there are in the world's flora. There may be as many as 800,000. Estimates for the angiosperms alone—the most conspicuous and most thoroughly studied elements of terrestrial vegetation—vary from the usually cited 200,000 to about half a million.[342]

A comparison of the number of plants valued as food and those employed as hallucinogens may be interesting. Of the vast assemblage of angiosperms, only about 3,000 species are known ever to have been used directly as human food. The number that actually feed the human race, however, is relatively very small, for only about 150 angiosperm species are important enough in human nutrition ever to have entered world commerce. Of these, only 12 or 13 species stand, in effect, between the world's population and starvation; these are all cultivated species.[349]

Very few plants provide the human race with narcotic agents, even though there may be many hundreds of species with psychoactive organic constituents. Nearly 5,000 species, for example, are now known to contain alkaloids, and most narcotics are of alkaloidal nature. It might, therefore, justifiably be expected that of these a'kaloid-containing plants alone, at least several hundred species might potentially be narcotic. Yet perhaps fewer than 75 species, including both cryptogams and phanerogams, are employed in primitive and advanced cultures as intoxicants. And of these 75, only about 20 may be considered of major importance. Furthermore, and perhaps significantly, only a few narcotic plants—coca, the opium poppy, hemp and tobacco—are numbered amongst the world's commercially important plants, and three of these (hemp excepted) are cultigens unknown in the wild state: an indication of long association with man and his agricultural practices.[342,346,349]

Of these 75 narcotics, the hallucinogens comprise only some 60 to 65 species, of which only a few, 10 or fewer, are cultivated; even then, with the exception of *Cannabis*, it is on a very primitive scale. Some of these hallucinogens (e.g. the tree-daturas of South America) are cultigens unknown in the wild and undoubtedly have long been associated with man and his magico-religious rites.[349]

While, as indicated above, there are many species of psychotropic plants that have never been employed as hallucinogens, it is true that there have been few cultures, even in the most restricted and limited floras, that have not ingeniously discovered and cleverly utilized at least one plant for its psychotropic activity. Lewin has appraised this interesting observation as follows: "The passionate desire which . . . leads man to flee from the monotony of everyday life . . . has made him instinctively discover strange substances. He has done so, even where nature has been most niggardly in producing them and where the products seem very far from possessing the properties which would enable him to satisfy this desire." [218]

Most interesting and possibly significant is the curious disparity between the number of species used as narcotics—especially as hallucinogens—in the New and Old World cultures. The explanation of this condition is far from clear as to whether it be due to cultural differences, to floristic peculiarities or to some other as yet unappreciated or recondite cause. Since there seems to be no botanical reason to presume that the flora of one hemisphere is richer in hallucinogenic species than that of the other, the explanation must more probably be sought in basic cultural and historical differences in the role of the hallucinogenic experience in the magico-religious aspects of primitive life in the Old and the New World.[210,334,335,338,348]

The plants used as hallucinogens are not evenly distributed throughout the plant world. Of the recognized divisions of the Plant Kingdom—the bacteria, fungi, algae (including diatoms), lichens, bryophytes, pteridophytes, gymnosperms and angiosperms—man employs as hallucinogens members only of the fungi and angiosperms. Notwithstanding the presence amongst the bacteria and algae of highly toxic or otherwise biodynamic constituents, these groups are, for the roster of hallucinogens, conspicuous because of their absence. Again, a clear and acceptable explanation of this unexpectedly disparate botanical distribution of man's hallucinogens has not yet been offered.[347]

Hallucinogenic constituents must be rather widely distributed throughout the fungi, but those species that man employs are concentrated almost exclusively in Basidiomycetes, where some genera of mushrooms and several species of puffballs have been reported. The Basidiomycetes represent the higher, evolutionarily most advanced fungi. There are definite hallucinogens, however, amongst the simpler and probably less advanced fungi, but they are not purposefully employed for their psychotomimetic properties. The best known is the Ascomycete, *Claviceps purpurea*, ergot, a rhizomorph that parasitizes the grains of rye and other grasses and which, when accidentally ground up into flour and unwittingly used in bread-making, has poisoned, often fatally, the population of whole areas in Europe before legislation and routine inspection corrected the cause. This poisoning, known as ergotism and, in the Middle Ages, as "St. Anthony's Fire," is characterized amongst many other effects by hallucinations; to the best of our knowledge, ergot has never been utilized primarily as an hallucinogen.

In the angiosperms, hallucinogenic plants are widely distributed, especially amongst the dicotyledons. In fact, only several hallucinogens can be listed—and these only on the basis of vague reports—for the monocotyledons, in spite of the high incidence of alkaloids in some of the families in this sector of the angiosperms. Thus, by far the greatest concentration of psychotomimetics used by man occurs in the dicotyledons, where they are represented in at least 30 genera of some 17 families. Ten of these families, or 59 percent, belong in the archichlamydeous part of the dicotyledons; the remaining seven are in the metachlamydeous assemblage. Curiously, only a small percentage of the genera of hallucinogens are to be found in dicotyledonary families known to be exceptionally rich in alkaloids: Leguminosae, perhaps five genera; Apocynaceae, only one; Solanaceae, six or seven. Some hallucinogens occur even in families which, until recently, were thought to be devoid of or very poor in alkaloids; for example, Myristicaceae, Convolvulaceae, Labiatae, Compositae. It is obvious that no broad chemotaxonomic significance whatsoever can be attached to the distribution of the dicotyledonary hallucinogens. It is, however, almost always true that, as stated by Farnsworth, ". . . When different genera or species of a particular plant family contain true psychotogens, these substances are always chemically similar, if not identical. This is a remarkable finding; chemotaxonomic relationships are not always so clear-cut." [99]

Chapter III

THE CHEMICAL DISTRIBUTION
OF HALLUCINOGENS

CONSIDERING the enormous number of different structures known in organic chemistry, it is interesting and possibly significant that psychotomimetic compounds are found within very few structural types.[165,347] These are listed in Table I.

Most of the hallucinogens are nitrogen-containing compounds, which means that they belong to the large and heterogeneous class of plant constituents known as alkaloids. Only one important type of hallucinogen of ten is non-nitrogenous.

When the structural types in Table I are compared, it is striking to see how often indole structures appear, and always in the form of tryptamine derivatives. These may be tryptamines with-

TABLE I
STRUCTURAL TYPES OF PRINCIPAL HALLUCINOGENS

A. Nitrogen Containing (Alkaloidal) Hallucinogens

I. Phenylethylamine Derivatives

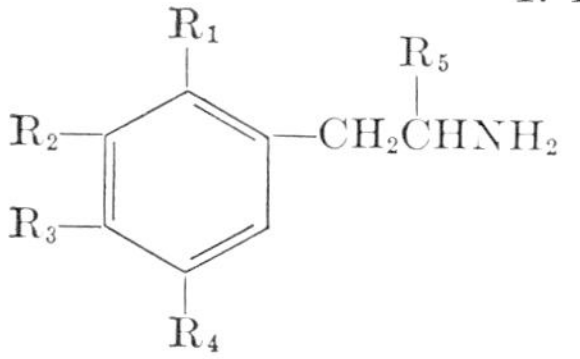

$R_1=R_5=H$, $R_2=R_3=R_4=OCH_3$: 3,4,5-Trimethoxy-phenylethylamine (constituent of peyote) = Mescaline

$R_1=R_4=OCH_3$, $R_2=H$, $R_3=R_5=CH_3$: 2,5-Dimethoxy-4-methyl-phenylisopropylamine = STP (synthetic compound)

II. Indole Derivatives

1. Tryptamine Derivatives

CH$_2$CH$_2$N(Alkyl)$_2$

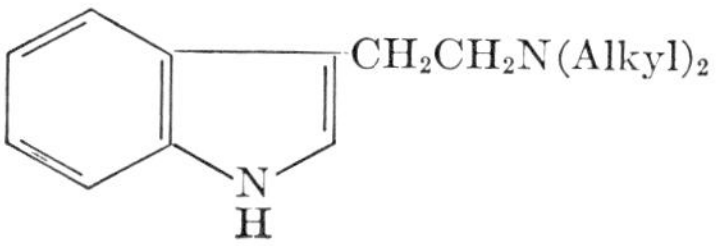

a) Alkyl$=$CH$_3$: N,N-Dimethyltryptamine (constituent of yopo, etc.)
b) Alkyl$=$C$_2$H$_5$, C$_3$H$_7$, C$_3$H$_5$: N,N,-Diethyltryptamine, etc. (synthetic compounds)

17

TABLE 1 (Continued)

2. Hydroxytryptamine Derivatives

  1) 4-Hydroxytryptamine Derivatives:

a) R=OPO$_3$H, Alkyl=CH$_3$: Psilocybin
R=OH, Alkyl=CH$_3$: Psilocin (constituents of teonanacatl)
b) R=OPO$_3$H, Alkyl=C$_2$H$_5$: CY-19
R=OH, Alkyl=C$_2$H$_5$: CZ-74
(synthetic compounds)

  2) 5-Hydroxytryptamine Derivatives:

a) R$_1$ = OH, R$_2$ = R$_3$ = CH$_3$: Bufotenin

b) R$_1$ = OCH$_3$, R$_2$ = R$_3$ = CH$_3$:
5-Methoxy-N,N-dimethyltrypta-
mine

c) R$_1$ = OCH$_3$, R$_2$ = H, R$_3$ = CH$_3$:
5-Methoxy-N-methyltryptamine
(constituents of yopo, cohoba,
epena, etc.)

3. Cyclic Tryptamine Derivatives

  1) β-Carboline Derivatives:

a) Harmine
b) 3,4-Dihydroharmine=Harmaline
c) d-1,2,3,4-Tetrahydroharmine
(constituents of ayahuasca, etc.)

  2) Lysergic Acid Derivatives:

a) R$_1$=R$_2$=H: d-Lysergic acid amide
b) R$_1$=H, R$_2$=CHOHCH$_3$: d-Lysergic acid hydroxy-
ethylamide (constituents of ololiuqui)
c) R$_1$=R$_2$=C$_2$H$_5$: d-Lysergic acid diethylamide
(=LSD-25) (synthetic compound)

  3) Ibogaine:

(constituent of *Tabernanthe iboga*)

relatives, the basidia mature in closed basidiocarps, and spore dispersal is effected by wind, water, insects or other animals. There are four orders in this series. The hallucinogenic members considered in this discussion belong to the order Lycoperdales and in the family Lycoperdaceae—the puffballs.

In the Hymenomycetes, the basidia do not mature in closed basidiocarps and spore dissemination through basidiospore abstriction and wind dispersal is highly efficient. There are two orders in this series. The hallucinogenic members are included in the order Agaricales—the gill fungi or mushrooms and one group of pore fungi; they are classically placed in the broadly defined family Agaricaceae.

Lycoperdales

The Lycoperdales, or puffballs, are cosmopolitan in distribution, with an estimated seven genera and in excess of 100 species; most of these species belong to the genus *Lycoperdon*. The basidiocarps, which are completely closed and usually hypogeous, consist of an outer wall or peridium and a spongy inner flesh or gleba. The basidia are borne within the gleba which breaks down upon maturation of the basidia, leaving only the basidiospores and a mass of capillitial threads in the peridium. The basidiocarp dries out to a tough skin filled with many millions of spores which are ejected through a pore at the apex of the basiciocarp by any bellowslike pressure exerted naturally or by animals.

Lycoperdaceae

Lycoperdon Linnaeus

There are an estimated 50 to nearly 100 species of *Lycoperdon*, native mostly to the temperate zone in moss-covered forests.

A recent report indicates that two species of puffballs—*L. marginatum* Vitt. and *L. mixtecorum* Heim—are employed as hallucinogens by the Mixtec Indians south of Tlaxiaco, in Oaxaca, Mexico, at an altitude of 6,000 feet or higher.

The Mixtecs call *L. mixtecorum*, an endemic of Oaxaca, *gi'-i-wa* ("fungus of first quality") and *L. marginatum*, a species known from temperate Europe and America, which has a strong odour of excrement, *gi'-i-sa-wa* ("fungus of second quality"). These two puffballs do not appear to occupy the place as divinatory agents that the mushrooms hold amongst other Indians of Oaxaca.[144,282]

The more active of the two species—*L. mixtecorum*—causes a state of half-sleep one-half hour after ingestion of one or two specimens. Voices and echoes are heard, and voices are said to respond to questions posed to them. One Indian reported: "I fell asleep for an hour or an hour and a half and the puffball spoke to me then, saying that I would become ill but would recover from the sickness." The effects of the puffball differ strongly from those of the hallucinogenic mushrooms; they may not induce visions, although definite auditory hallucinations are reported to characterize the intoxication.

Narcotic effects of puffballs have been recorded in the literature, although these reports from Mexico constitute apparently the first that indicate the deliberate utilization of the plant for intoxication. From the southeastern United States, a report, dating from 1869, states:

> It has been mentioned by medical writers that the spores of the puffballs have narcotic properties, and it is an analgesic agent, acting somewhat like chloroform when inhaled, but I have never experienced any effects of the kind from its use as a vegetable. However, Dr. Harry Hammond . . . writes to me, "since writing to you, I and a number of others have made several meals on *Lycoperdon*, and I think I have discovered in myself well-marked evidences of a narcotic influence—and two other experimenters have described similar sensations to me. I recollect also to have heard from Mr. Mahan that a friend of his, a physician in Georgia, had been seriously affected in this way by too large a meal on *Lycoperdon*.[144]

There is as yet no phytochemical basis on which to explain the intoxicating effects of these two Gasteromycetes.

Lycoperdon mixtecorum *Heim,* Comptes Rend, 254
(1962) 789; Rev. Mycol, 31 (1966) 156.

Receptacle 2–3 cm in diameter, subglobose, slightly flattened, abruptly constricted into short peduncle 3–4 mm long. Exoperidium not echinulate but densely cobbled-pustuliform, light tan.

Endoperidium silky, papyraceous, smooth, straw-coloured. Peridial envelope yellowish brown mixed with orange, covered with a whitish chevelure at base. Opercule ragged. Gleba loosely spongy, grey-tawny to slightly violet, capillitium filaments straight, irregular, 2–6μ long. Sterile base slightly developed, lemon-yellow to nearly orange, cellules relatively large (2–3 mm) radially oriented. Spores brownish tawny with subtle violet tinge,

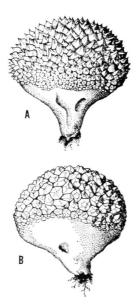

Figure 5. *Lycoperdon marginatum* (A) and *L. mixtecorum* (B). From R. Heim et al. *Nouvelles Investigations sur les Champignons Hallucinogènes*. Editions du Muséum National d'Histoire Naturelle, Paris. (1967) 196.

spherical 7.8–10μ including sculpturing, muricate-winged, presenting 5 distinct membranes covered with a mesh of incomplete unequal pale threads. Terrestrial in light forest and in pastures. Oaxaca, Mexico.

Agaricales

The Agaricales, containing the most familiar of the fungi— the mushrooms—is credited by one modern systematist with 16 families and 197 to 200 genera, by another with 7 families, about 275 genera and 7000 species. Members of this order are saprophytes or parasites with a well developed, sometimes perennial mycelium hidden in the substratum (soil, rotting wood, etc.). The reproductive structure or fruiting body is usually macroscopic, showing great morphological diversity, and has a well marked hymenium which likewise varies appreciably, the differences often providing characters to distinguish families. All species are holobasidial with the usually simple, clavate basidia persistently unicellular (with apical sterigmata and 4 basidiospores) and generally standing side by side with their longitudinal parallel to the same axis of neighbouring basidia providing a kind of palisade—the hymenium.

Agaricaceae

All of the genera of intoxicating mushrooms considered in the following discussion—*Amanita, Conocybe, Panaeolus, Psilocybe, Stropharia*—have classically been included in one broad family: the Agaricaceae. There are good reasons—at least in the present work—to continue using the older classification, since many of the characters employed in the newer systems are recondite and often do not outline differences of sharp delineation. These family concepts are based often on differences in morphology of the hymenophoral trama and on spore type and colour in addition to numerous other more minor characters.

The Agaricaceae, in its broad sense, includes the gill-bearing fungi which are distinguished by having the fruiting area on platelike folds (gills or lamellae) usually on the under surface of the pileus. The reproductive bodies are generally carnose, and, although they vary considerably, are usually pileiform or umbrella-shaped, with a cap or pileus and a stipe. When young, the reproductive body may be enclosed in a veil which ruptures with growth, often persisting as a volva at the base of the stipe and, in patches, on the top of the pileus. The species are mostly terrestrial, occurring most frequently in loam, sandy soil or humus in forest habitats, often in moss, occasionally in pastures and on lawns, sometimes on living or decaying plant material. The family is almost cosmopolitan in its range, but the species are usually of limited distribution.

Amanita Linnaeus

The genus *Amanita* comprises some 50 to 60 species (although some estimates reach 100). It is almost cosmopolitan, occurring on all continents except South America and Australia, but the individual species are usually restricted to definitive areas. A number of the species are variously toxic, and their chemical constitution, although in general rather poorly known, seems to be rather variable; the toxic principles are not the same in all poisonous species. Some species are edible and have been famous as foods in Europe since classical times; a few species have antibiotic properties.

Some modern treatments place *Amanita* in the family Amanitaceae, separated from the closely related Agaricaceae on technical characters of the morphology of the hymenophoral trama, type of spores and other minor characters.

Probably the oldest of the hallucinogenic plants and once perhaps the most widespread (insofar as man's utilization of it is concerned) may be the *fly agaric*, a mushroom of the north temperate zone of Eurasia: *A. muscaria*. This species occurs also in temperate North America, but it is represented apparently by a different chemical variant in the New World and never has been employed narcotically as in the Old World. In spite of the great age of its use, only recently has a clarification of the chemistry of its active principles begun to take shape.

Europeans discovered the narcotic use of *A. muscaria* amongst primitive tribesmen of Siberia in the eighteenth century, when, in 1730, a Swedish army officer published a book on his 12 years as a Russian prisoner in Siberia. Until very recently, it was employed as an orgiastic or shamanistic inebriant by scattered groups—the Ostyak and Vogul, Finno-Ugrian peoples in western Siberia; the Chukchi, Koryak and Kamchadal of northeastern Siberia along the Pacific coast. Tradition has also established its use amongst other groups in this vast region. It has even been

.I.

Amanita muscaria Fries ex Linnaeus

The "fly agaric," used as an
hallucinogenic agent in Siberia.

Drawn from Heim: *Champignons toxiques et hallucinogènes*

Figure 6. Drawn by I. Brady, after illustration from R. Heim. Editions
N. Boubée and Cie, Paris (1963).

Two other pathways for the synthesis of ibotenic acid have been suggested by Sirakawa and others,[365] and by Kishida and others.[200]

Besides the preparation of muscimole by decarboxylation of ibotenic acid, two different pathways for synthesis of muscimole have been described.[110]

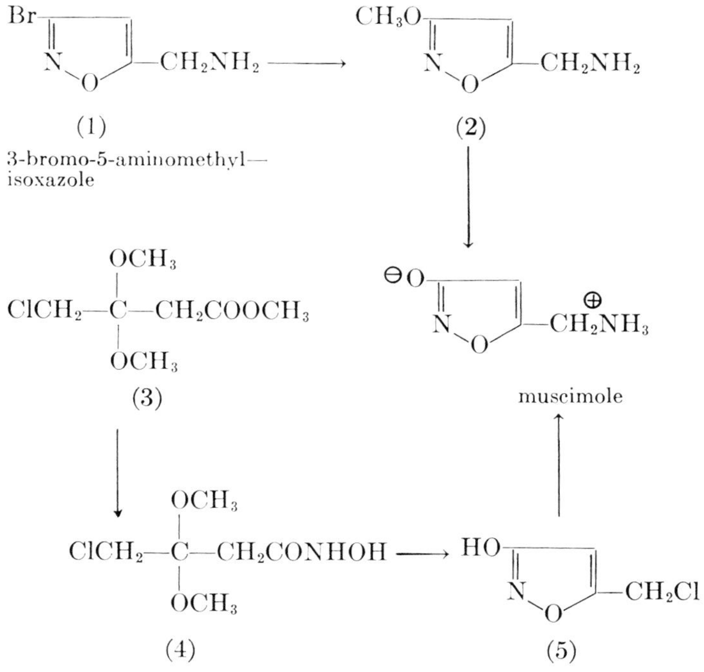

Br⌐‖⌐
N‿O⌐—CH$_2$NH$_2$ ⟶ CH$_3$O⌐‖⌐
N‿O⌐—CH$_2$NH$_2$

(1)   (2)

3-bromo-5-aminomethyl—
isoxazole

OCH$_3$
|
ClCH$_2$—C—CH$_2$COOCH$_3$
|
OCH$_3$

(3)

⊖O⌐‖⌐
N‿O⌐—CH$_2$NH$_3$⊕

muscimole

OCH$_3$
|
ClCH$_2$—C—CH$_2$CONHOH ⟶ HO⌐‖⌐
N‿O⌐—CH$_2$Cl
|
OCH$_3$

(4)   (5)

One synthesis starts with 3-bromo-5-aminomethyl-isoxazole (1), which, by heating with KOH/MeOH, is transformed into 3-methoxy-5-aminomethyl-isoxazole (2). This compound, after hydrolysis, yields muscimole.

Another method utilizes as starting material the ketal of γ-chloro-acetoacetate (3), which is treated with hydroxylamine to provide the corresponding hydroxamic acid (4). Cyclization with dry HCl gas in absolute acetic acid affords 3-hydroxy-5-chloro-methyl-isoxazole (5), which can be transformed by treatment with NH$_3$ into muscimole.

Studies with ibotenic acid and with muscimole in pharmacology and experimental psychology have shown that there is no significant qualitative difference between these two substances; however, quantitatively muscimole proves to be at least five times more active than ibotenic acid.

In pharmacological experiments on animals, the principal demonstrable effect is inhibition of motor functions. This is brought about by a central nervous supraspinal mechanism of action. Vegetative functions, however, are hardly influenced by these two substances.

Psychological experiments with normal test persons showed that both ibotenic acid and muscimole cause a relatively uncharacteristic condition of intoxication.

Muscimole, with a 10–15 mg oral dose in man, creates such symptoms as confusion, disorientation in situation and time perception, disturbance of visual function and hearing, muscle twitching, weariness, fatigue and sleep. Psychic performance and learning are diminished. The effects simulate a type of acute exogenic reaction but do not lead to a genuine model psychosis, such as can be produced with LSD or psilocybin.[399, 420]

Two additional constituents of *A. muscaria*, only very recently isolated, which may contribute to the pharmacological activity of this mushroom are another amino acid derivative, (−)-R-4-hydroxy-pyrrolidone-(2), colourless crystals, mp 153-155°[α]$_D^{23}$ = −44,5° (methanol), and an indolic compound, (−)-1,2,3,4-tetrahydro-1-methyl-β-carboline-carboxylic acid.[232]

The hydroxy-pyrrolidone possesses narcotic-antagonizing activity. Up to the present time, no results concerning the pharmacological properties of the indolic compound can be reported.[91]

Figure 7. *Amanita muscaria*, the "fly agaric." Switzerland. Photograph by C. H. Eugster.

Amanita muscaria (*L. ex. Fr.*) *Persoon ex W. J. Hooker*
Fl. Scot., (May 1821) 19.

Pileus 7–20 cm in diameter, oval, becoming hemispheric to convex and finally plane, viscid when immature, whitish to yellow or, usually in Eurasia, orange to scarlet, normally speckled with rough, yellow or white warts, marginally striate, flesh white, yellow under pellicle. Gills white or cream, free or attached to stipe with short line. Stipe up to 20 cm tall, 1–2.5 cm in diameter, cylindrical or sometimes tapering up, hollow or filled with a fibrillate mass, white or cream, basally bulbous and covered with encircling scales. Ring white or cream, ample, usually membranaceous, persistent. Volva floccose, remaining as a yellowish lacerate collarlike ring or scale. Spores white, ovate, apiculate, 7.5–10 × 6.3–7.5μ. In deciduous or coniferous forests and overgrown pastures. North temperate zone of both hemispheres, occurring up to 6,000 feet altitude, in the Old World ranging south to North Africa, where it grows in oak and eucalyptus groves.

There are many "forms" of *A. muscaria*, recognizable by their shape, size and colour of the pileus. There appear also to be significant chemical differences between the Eurasian and North American forms.

Conocybe Fayod; *Panaeolus* (Fr.) Quélet; *Psilocybe* (Fr.) Quélet; *Stropharia* (Fr.) Quélet

Conocybe, a genus of some 40 species that grow in forests and pastures and gardens on dung, charcoal, sandy soil, ant hills and decayed wood, is of cosmopolitan range. While it can be accommodated in the Agaricaceae as classically delimited, Singer places it in a separate family, the Bolbitiaceae.

Panaeolus, a cosmopolitan genus of some 20 species, is included in the Agaricaceae in the broad sense but in the Coprinaceae by Singer, the Strophariaceae by Heim. The species occur typically on soil and dung.

Psilocybe, undoubtedly the most important genus of the "sacred" Mexican hallucinogens, represents an almost cosmopolitan genus. It is found from the arctic to the tropics, although its main distribution is temperate. The species grow in the soil and on a variety of organic substrata such as humus, dung, rotting wood, bagasse, peat and also in clumps of mosses. Included classically in Agaricaceae, some modern systematists place it in Strophariaceae. About 40 species are now recognized, but there are undoubtedly many more to be described. Most of the hallucinogenic species fall into section *Caerulescentes* of the genus.

Stropharia, customarily put into Agaricaceae in the past, is now often assigned to Strophariaceae. Singer has sharply delimited the generic boundaries and accepts as *Stropharia* only nine species, whereas earlier systematists recognized some 70 specific concepts. Its distribution is almost cosmopolitan. The species grow usually in the soil and on dung, but some occur on leaves and rotting wood.

The Spanish conquerors of Mexico were much disturbed by an important religious cult based on the sacramental consumption of sacred mushrooms called *teonanactl* ("flesh of the gods") by the Aztecs. The Indians communed with the spirit world through the mushroom-induced hallucinations. Divination, prophecy and curing rites likewise depended on the narcotic effects of these fungi.[308,321,323,427,429,433]

As with other Mexican religions that utilized inebriating plants, European persecution drove the mushroom cult into hiding in the hinterlands. Most of the early chroniclers were clerics who emphasized the need for stamping out such a loathsome custom as the sacramental eating of toxic mushrooms. Their criticism of teonanacatl was even more vehement than criticism of the peyote cactus and *ololiuqui*, the morning glory. As mycophobes, their religious fanaticism was easily directed towards a despised form of plant life which, through its vision-inducing powers, held the Indian in awe. There was little that Christianity could offer comparable in the Indian mind to the supernatural power of the mushrooms.

The mushroom cult appears to have deep roots in centuries of native tradition. Frescoes from central Mexico, dated at 300 A.D., have drawings which would seem to put mushroom worship back at least 1700 years. Even more remarkable are the archaeological artifacts now called "mushroom stones," excavated in great numbers from highland Mayan sites in Guatemala. These are dated conservatively at 1000 B.C. They consist of an upright stipe with either a manlike or an animal figure crowned with an umbrella-shaped top. These stones were long a puzzle, and some archaeologists supposed them to be phallic symbols. Modern students, however, quite widely hold that they represent a kind of icon connected with mushroom worship. The real significance of these stones lies in the revealing of the existence of a sophisticated mushroom cult at such an early date and far beyond the present geographical limits of the magico-religious use of the fungi.[43,147,429]

One of the first European references to teonanacatl mentioned mushrooms "which are harmful and intoxicate like wine" so that those who partake of them "see visions, feel a faintness of heart and are provoked to lust." The natives were reported to eat the mushrooms with honey and "when they begin to be excited by them start dancing, singing, weeping. Some . . . see themselves dying in a vision; others see themselves being eaten by a wild beast, others imagine that they are capturing prisoners of war, that they are rich, that they possess many slaves, that they have committed adultery and were having their heads crushed for the offense." Other early reports detail the use and effects of teonanacatl similarly, and several offer crude illustrations of the mushrooms.[321]

Hernández, the King of Spain's personal physician, wrote that three kinds of narcotic mushrooms were worshipped by the Aztecs. Those called *teyhuintli* "cause not death but madness that on occasion is lasting. . . . There are others that . . . bring before the eyes all sorts of things, such as wars and the likeness of demons. Yet others there are not less desired by princes for their festivals and banquets, and these fetch a high price. With night-long vigils are they sought, awesome and terrifying. This kind is tawny and somewhat acrid." [150]

Despite the great age of the mushroom cult and the numerous detailed and forceful Spanish reports of this curious use of the fungi, our knowledge of their identification, utilization and chemistry is all recent. Although toxic mushrooms were known in Mexico, none had, in four centuries, been found employed for magico-religious purposes. Some 50 years ago, an attempt at identifying teonanacatl proposed that the "sacred mushroom"

search, in its effort to obtain novel compounds which are valuable in medicine, can revert to ancient knowledge of the miraculous powers hidden in the Plant Kingdom." [165]

The active principles of teonanacatl, the "sacred mushrooms" of Mexico, are psilocybin and psilocin, the former being the main component. These compounds were first isolated from *P. mexicana* in 1958 by Hofmann and others.[172]

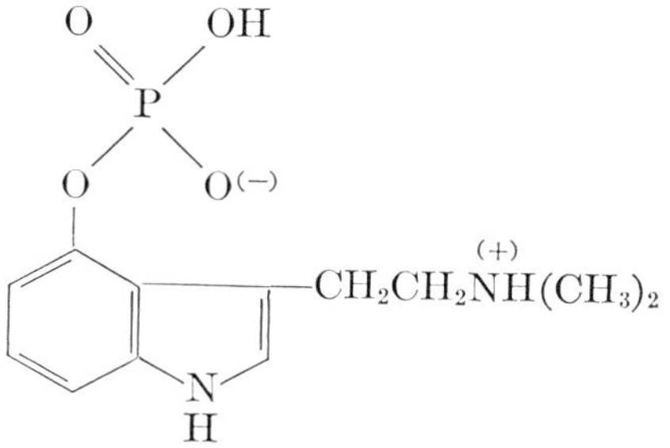

Psilocybin

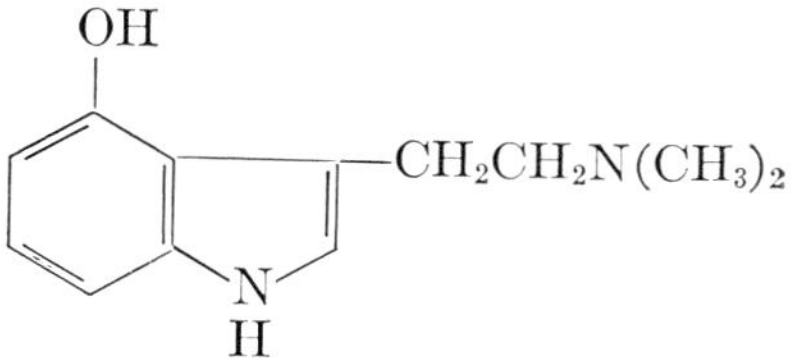

Psilocin

Because of the inconclusive results with animal tests in attempts to follow the active principles through the extraction and chromatographic procedures, human experiments had to be carried out at various stages during the concentration of the active principles.[157] Later, other *Psilocybe* species belonging to the teonanacatl group were also shown to contain psilocybin, usually together with a small amount of psilocin: for example, *P. caerulescens* var. *mazatecorum*, *P. zapotecorum*, *P. aztecorum*, *P. semperviva* as well as in *Stropharia cubensis*.[146] Furthermore, psilocybin and psilocin have also been found in North American species of *Psilocybe* which are not known to be used for psychotomimetic purposes: *P. pelliculosa*,[408] *P. cyanescens*, *P. baeocystis* and in the botanically closely related species *Conocybe cyanopus*[32] and in *P. quebecensis.*[253]

A European species of *Psilocybe, P. semilanceata,* was also found to contain psilocybin.[174] Psilocybin has been isolated from species of *Panaeolus:* for example, *P. foenisecii.*[250,251,296]

From artificial cultures of *Psilocybe baeocystis,* demethylated derivatives of psilocybin were isolated, and these principles were named baeocystin and norbaeocystin.[215]

Baeocystin   Norbaeocystin

Degradation studies showed psilocybin to be 4-phosphoryloxy-N,N-dimethyltryptamine. Hydrolysis of psilocybin gives equi-molecular amounts of phosphoric acid and psilocin, which is 4-hydroxy-N,N-dimethyltryptamine.[171,175] These structures were confirmed by synthesis.[173]

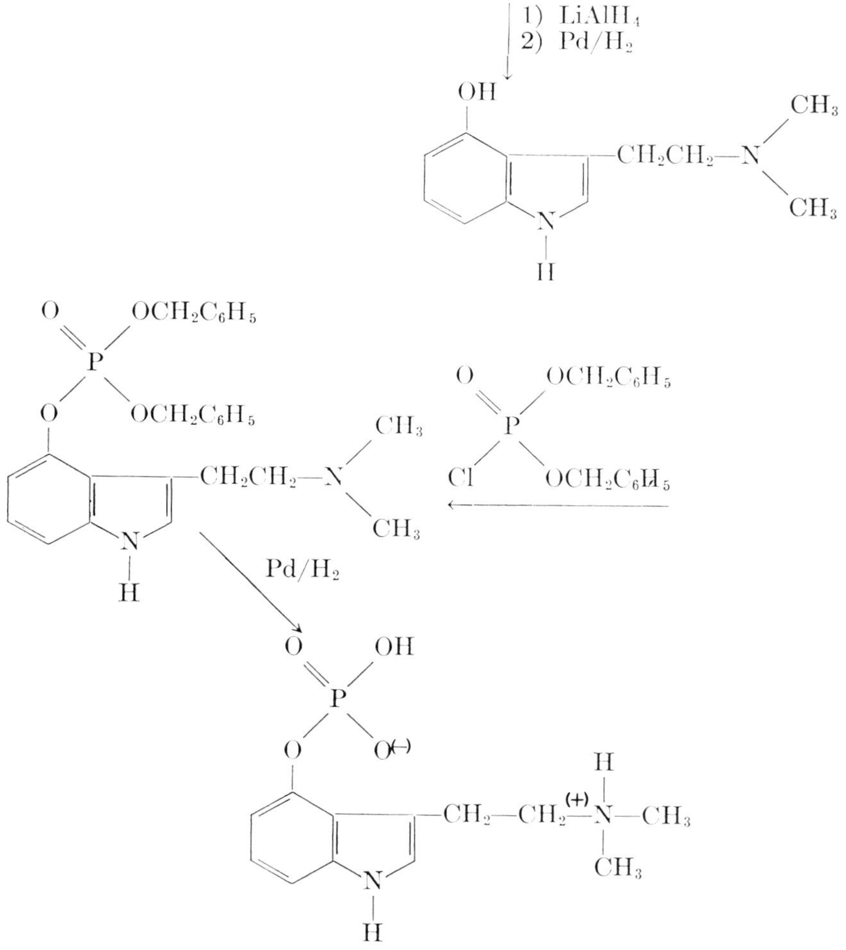

1) $LiAlH_4$
2) Pd/H_2

OH

$-CH_2CH_2-N$ $\begin{array}{c} CH_3 \\ CH_3 \end{array}$

N
H

O $=$ P, OCH$_2$C$_6$H$_5$, OCH$_2$C$_6$H$_5$, O

$-CH_2CH_2-N$ $\begin{array}{c} CH_3 \\ CH_3 \end{array}$

N
H

O $=$ P $\begin{array}{c} OCH_2C_6H_5 \\ OCH_2C_6H_5 \end{array}$ Cl

Pd/H_2

O $=$ P, OH, O, O$^{(-)}$

$-CH_2-CH_2-\overset{(+)}{N}-CH_3$
H
CH$_3$

N
H

Psilocybin

The synthetic production of psilocybin is a much more rational operation than the isolation of it from the mushrooms. The dried mushrooms contain 0.2 to 0.4 percent of psilocybin. Psilocin is present only in trace amounts. Psilocybin is a stable compound readily soluble in water and obtainable in colourless crystals. Psilocin, on the other hand, is very sensitive to oxidation and difficultly soluble in water. Psilocybin is the first and until its discovery the only known natural indole compound with a phos-

phoric acid radical. Psilocybin and psilocin are likewise novel in
that they are substituted by a hydroxy group in the 4-position of
the indole structure.

Tryptophane seems to be the biogenetic precursor of psilocybin.
D,L-tryptophane-[β-C^{14}] was utilized at a rate of 10 to 20 percent
by the fungus *Psilocybe semperviva* in producing psilocybin.[46]

The medium oral dose of psilocybin for man is 4 to 8 mg; it
elicits the same symptoms as the consumption of about 2 gm of
the dried mushroom, *P. mexicana,* according to reports of self-
experiments with the mushroom and with psilocybin by R. Heim,
A. Hofmann, A. Brack, and R. Cailleux.[147] The psychotomimetic
effects of psilocybin, psilocin, LSD, and mescaline were compared
by Wolbach and others,[446] and found to be qualitatively similar.
The time course of psilocybin and psilocin reactions is shorter than
that of LSD or mescaline reactions. Psilocin is approximately 1.4
times as potent as psilocybin. This ratio is the same as that of the
molecular weights of the two drugs. The development of "cross"
tolerance between LSD and psilocybin supports the idea that
these two drugs cause psychic disturbances by acting on some
common mechanisms or on mechanisms acting through a common
final pathway.[3,184,186]

Psilocybin does not exhibit typical effects on isolated organs
(intestine, uterus, heart), with the exception of a pronounced inhibiting effect towards serotonin. On the entire animal, however,
it has characteristic autonomic effects: dilation of the pupils,
contraction of the nicititating membrane, pilo-erection, temperature increase, and so on. This is an ergotropic excitation syndrome, which results mainly from a central stimulation of sympathetic structures.[61,435] A very characteristic effect of psilocybin is
the regular enhancement of monosynaptic spinal reflexes: for
example, the patellar reflex of cats.[436]

The toxicity of psilocybin in animals is very low in comparison
to the effective dose in man. The LD$_{50}$ for the mouse is 280 mg/kg;
that is, psilocybin is 2.5 times less toxic than mescaline in this
test, while it has a 50 times higher psychotomimetic effect in
man.[157]

From the large number of synthetic derivatives of psilocybin
and psilocin,[401] the N,N-diethyl analogues: 4-phosphoryloxy-N,N-

diethyltryptamine (CY-19) and 4-hydroxy-N,N-diethyltrypta-mine (CZ-74), were tested in man.[214] These two compounds do not show any significant quantitative or qualitative differences in their effect. They differ from psilocybin and psilocin in that their period of action is somewhat shorter, amounting to an average of 3½ hours.

Conocybe siliginoides Heim, Comptes Rend. 242 (1956) 1390.

Pileus 1.3–2.3 cm in diameter, 0.9–1.9 cm tall, at first subhemi-spherical, then conic-campanulate, never spread, fawn-orange-red, near centre slightly deeper orange, glabrous, dull becoming shiny, hygrophanous, marginally regularly crenulate, white with darker striations. Stipe slender, rigid, cylindrical (hardly swollen towards base), up to 6 cm tall, 0.15 cm in diameter, white-farinaceous to top, pale orange in upper part, cream-citrine else-where, tinged with dull pink near middle, at first darker, always white at base, fistulous; growth of stipe continuing after growth of pileus. Flesh thin, translucid in pileus, 11–15×7–10×6–12μ, white with a slight tinge of flesh colour. Gills more or less distant, rather thick, adnexed, accompanied by two series of very un-equal, saffron-coloured or brownish orange lamellules. Spores very polymorphic, obovoid, very slightly cylindrical, often subtly hexagonal in profile, with a large germinative pore, bright ocher-ous or chrome yellow.

Known from Oaxaca, southern Mexico, growing on rotting tree trunks; fruiting in June and July.

Panaeolus sphinctrinus (*Fr.*) Quélet, Les champignons du Jura et des Vogues, pt. 1 (1872) 151.

Pileus 2–5 cm in diameter, at first conic or subovoid to cam-panulate, sometimes with central prominence, brown-grey or olive-grey, glabrous, hydrophanous, often somewhat areolate in dry state, margin at first slightly incurved, appendiculate from fragments of veil. Flesh thin, in colour similar to surface, more or less odourless. Stipe hollow, 6–12.5 cm long, 0.2–0.3 cm in diam-eter, equal, greyish to reddish brown, apically striate, basally somewhat swollen. Gills ascending-adnate, seceding, subdistant, broad, greyish at first becoming blackish and mottled, edges white-flocculose. Spores black, lemon-shaped, 13–19 × 9–12μ, smooth.

A widespread species known from many temperate parts of the world; growing in small groups in forests, pastures, fields and along roadsides, almost always in cow dung; fruiting in summer and autumn.

Figure 9. *Panaeolus sphinctrinus.* Courtesy of G.-M. Ola'h.

Psilocybe mexicana Heim, Comptes Rend. 242 (1956)
967; *Rev. Mycol.* 22 (1957) 77.

Pileus 0.5–2 (rarely up to 3) cm in diameter, 0.4–1.9 cm high,
conic-campanulate, sometimes becoming hemispherical to convex, often with central papilla, sometimes broadly conic becoming umbonate, umbo often with a bud-shaped papilla, very
hygrophanous, glabrous or glabrescent; margin at times sulcate-
striate to an apical disk; disk deep ocherous to ocherous brown,
brownish red, nearly lilac-hued in growing state then fulvous,
marginal area paler, grey-brown; silky white veil soon disappear-

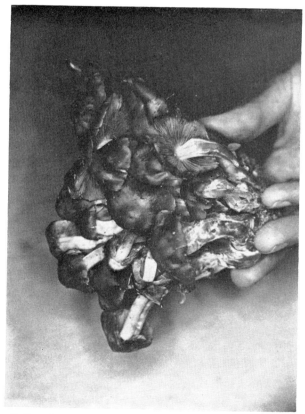

Figure 10. *Psilocybe aztecorum*. Oaxaca, Mexico. Photograph by R. G.
Wasson.

ing from margin. Stipe 2–6 (rarely up to 8) cm long, hollow, gracefully tapering upwards, often flexuous especially in lower part which is nonbulbous, ocherous to lightly yellowish pink, paler above, fulvous or slightly reddish in central portion, red-brown or grey-brown basally; veil membranaceous, leaving appressed silky fibrils on upper third, evanescent. Flesh in pileus paler than on surface, pale yellow, cream-pink in stipe, not cyanaceous, but turning bluish on bruising, odour farinaceous. Spores about $8–12 \times 5–8 \times 5–6.7\mu$, compressed, obovoid, sub-

Figure 11. *Psilocybe mexicana.* Oaxaca, Mexico. Photograph by R. G. Wasson.

Figure 12. *Psilocybe yungensis*. Photograph by R. G. Wasson.

isodiametric, smooth. Spore deposit deep sepia to dark purple-brown.

Known from southern Mexico and Guatemala between 4500 and 5500 feet, especially in limestone regions, growing isolated or sparsely in moss or herbs along roadsides, humid meadows, cornfields and in the neighbourhood of pine and oak forests; fruiting from May until October.

Stropharia cubensis Earle, Est. Agron. Cuba 1 (1906) 240.

Pileus 1.6–5 (rarely 8.5–12.5) cm in diameter, conic-campanulate, at first papillose at apex, then becoming convex to plane, sometimes with depression around umbo, umbo becoming obtuse (umbo or papilla sometimes absent), chocolate-brown or brown-orange, umbo yellow, pale tan to whitish near periphery, central area more or less fulvous, usually cyanaceous in age or upon injury, clearly viscous with sparse, distant floccose scales, becoming shiny, glabrous; margin even, entire, not appendiculate. Stipe 4–7 (sometimes up to 15) cm long, 0.4–2 cm in diameter, hollow, usually thickened near base, stiff, often bent, white becoming yellowish or ashy red, strongly sulcate-striate, otherwise smooth, glabrous to filbrillose, not hygrophanous; ring ample, membranaceous, smooth, fragile, irregular, ragged or lacerate, whitish becoming purplish black. Gills entire to undulate, serrate, adnate to adnexed, narrow, ventricose in mid-portion whitish to deep greyish violet or purple-brown, somewhat mottled. Flesh white, cyanescent upon injury, odourless. Spores ellipsoid, smooth, opaque, purple-brown, $11.5–17.5 \times 8–11.5 \times 7–9\mu$.

Known from southern Mexico, Central America, West Indies, Florida and reported from South America and southeastern Asia. Growing singly or in small groups, usually on dung or rich pasture soil, bagasse; fruiting from February to November or December, south of the equator from November to April. Some systematists recognize three "colour varieties."

THE ANGIOSPERMAE

With the exception of those included in the Fungi, all known or suspected hallucinogenic plants belong to the Angiospermae.

Together with the Gymnospermae, the Angiospermae constitute the Phanerogamae—embryo-producing plants in which fertilization is brought about by means of a pollen-tube. Differing from the Angiospermae basically in having naked seeds borne in open sporophylls in "cones," the Gymnospermae, a small group of some 700 living species, are not known to possess hallucinogenic constituents. The Angiospermae, the dominant group of land plants today with at least 300,000 (but probably many more) species, have the ovules protected by ovarian tissue and borne in closed sporophylls or carpels in typical "flowers."

The Angiospermae are subdivided into two divisions or classes: the Monocotyledonae, in which the embryo has one cotyledon or seed leaf; and the Dicotyledonae, with two (rarely more) cotyledons. The former produce no true wood, have leaves predominantly parallel-veined and flowers constructed usually in three parts. The latter often produce true wood, have leaves generally pinnately and reticulately veined and flowers predominantly four- or five-parted. The Monocotyledonae comprise approximately one quarter to one fifth as many species as the Dicotyledonae. The latter group is far more diverse than the former. Several presumed hallucinogens have been reported from the Monocotyledonae, but by far the greater number are included in the Dicotyledonae.

The Dicotyledonae are further subdivided into the Archichlamydeae and the Metachlamydeae, on the basis of characters of the corolla. The former, usually considered evolutionarily the more primitive, has simple, undifferentiated perianths with the petals, when present, separate: that is, it comprises all polypetalous and apetalous families. The latter, usually considered more advanced, has the petals united into a more or less tubular or bell-shaped, lobed organ called the corolla tube.

Hallucinogenic plants are represented in both subclasses of the Dicotyledonae, and, except within very broad limits or in isolated instances, there does not appear to be chemotaxonomic significance to the distribution of the hallucinogenic constituents.

CANNABACEAE
Dicotyledonae

The Cannabaceae (sometimes erroneously rendered Cannabinaceae or Cannabidaceae), belonging to the order Urticales, comprises two genera, *Cannabis* and *Humulus;* native to temperate parts of both hemispheres, it is now almost cosmopolitanly distributed, albeit less commonly in the wet tropics. Very closely related to the Urticaceae and Moraceae, with both families it shares a number of common characters and it is often placed in the Moraceae as a section or subgenus. There appear, however, to be valid morphological and chemical reasons for maintaining Cannabaceae as distinct.

Cannabis Linnaeus

Cannabis is usually considered to be monotypic but some botanists maintain that it comprises several species. Often included in the Moraceae or even the Urticaceae, it is now more commonly set apart, together with the genus of the hops plant, *Humulus*, in a distinct family: the Cannabaceae. *Cannabis* is the source of hempen fibre, an edible seed-oil, and of various narcotic preparations. The genus is now thought to be native to central Asia.

Undoubtedly, one of the oldest psychotomimetics used by man and today the hallucinogenic species most widely disseminated around the world is *Cannabis sativa.*

Figure 13. *Cannabis sativa*. Minneapolis, Minnesota. Photograph by R. F. Raffauf.

One of man's most ancient cultigens, hemp is known to have been valued by the Chinese 8,500 years ago. The Assyrians used the plant in the ninth century B.C. in the form of an incense. The Sanskrit *Zend-Avesta* first mentioned its intoxicating resin in 600 B.C. Herodotus wrote that the Scythians burned its seeds to produce a narcotic smoke. In Thebes, it was made into a drink said to possess opiumlike properties. Galen recorded general use of hemp in cakes which, if eaten to excess, had intoxicating properties. In thirteenth century Asia Minor, the hashishins, who were political murderers, were promised, as reward for work well done, supplies of hasheesh—the resin from *Cannabis;* from this association comes to European languages, through Arabic, the modern word *assassin*.[189,218,349]

Although the narcotic use of *Cannabis* harkens back thousands of years in India, the Near East, parts of Africa and other areas of the Old World, its spread to nearly all inhabited parts of the globe has allowed its employment as an inebriant to increase recently in sophisticated societies, especially in urban centres, and to lead to major problems and dilemmas for European and American authorities. Studies in depth of its utilization in less developed societies and modern improvement of chemical techniques should shed much light on some of the problems and dangers resulting from its use and abuse in more advanced communities.

Methods of using *Cannabis* vary widely. In the New World, *marijuana* (or, in Brazil, *maconha*)—the dried, crushed flowering tops and leaves—are almost invariably smoked, often mixed with tobacco, in the form of cigarettes. In parts of primitive Africa, *Cannabis* fulfills an important role in religion and magic, bespeaking very long usage. In southern Africa, it is called *dagga*, a term sometimes also applied, albeit with a qualifying adjective, to sundry species of the labiate genus *Leonotis*, several species of which have feeble narcotic effects when the leaves are smoked. In Morocco, where the use of *Cannabis* is common, the vernacular name is *kif*. Hasheesh, the resin from the recently fertilized pistillate flowers, is smoked, eaten or drunk by millions, especially in Moslem areas of North Africa and the Near East.[218]

It is, however, in India where apparently *Cannabis* assumes an important religious significance in certain cults and where, as a result, man has selected "races" characterized by high concentrations of tetrahydrocannabinol. The ancient Indian *Atharva-Veda* called the drug a "liberator of sin" and "heavenly guide." The plant is still held sacred in many temples, where it is planted out in gardens. Indians commonly employ three *Cannabis* preparations as narcotics. *Bhang*, the weakest, consists of the dried plant gathered green, powdered and made into a drink with water or milk, or, with sugar and spices, into candies called *majun;* opium and *Datura* are said sometimes to be added. *Ganja*, usually smoked with tobacco but sometimes eaten or drunk as an infusion, consists of dried pistillate tops with exuded resin carefully removed from cultivated or escaped "races" notably rich in tetrahydrocannabinol. *Charas*, pure resin removed from leaves and stems also from especially cultivated, strongly narcotic "races," is normally smoked, but it may also be eaten mixed with spices. *Cannabis* is the source of the narcotic for the poor in India, where, in addition to its conspicuous religious use, it is highly valued in folk medicine and as a presumed aphrodisiac. Hedonistically, it is likewise valued as an euphoric narcotic, especially in activities requiring endurance or physical effort.[218]

Even though the marked increase in the smoking of marijuana in the United States and Europe poses a variety of complex problems, much of the drug illicitly used at the present time in these north temperate areas is weak, sometimes probably almost lacking, in the narcotic principles since it consists not of pure resin but of crushed leaves, twigs and tops of plants which may be low in the tetrahydrocannabinols. These plants grow spontaneously, often having spread mainly from hemp formerly cultivated in plantations for fibre production. In the United States, where the hemp-fibre industry has all but disappeared, the cultivation of *C. sativa* was once a major agricultural industry. Marijuana smuggled into the United States from Mexico and other hot, dry regions, or hasheesh introduced into Europe from the Near East and Africa represent a stronger and, consequently, a potentially more troublesome and pernicious narcotic.[349]

In spite of its great age as one of man's principal narcotics and its utilization by millions of people in many cultures, and notwithstanding the great economic value of the plant for uses other than as an intoxicant, *Cannabis* is still characterized more by what we do not know botanically about it than what we know.

Lack of knowledge about *Cannabis* and its utilization as a narcotic not only provides an obstacle to an understanding of moral, legal, sociological and economic phases of its importance to the cultures where its use has become established, but even many technical aspects—botanical, chemical, pharmacological, medical and public health—are fraught with contradictions and uncertainties.

Botanists have long tended to believe *Cannabis* to be monotypic and that its one polymorphic species, *C. sativa*, has diversified into many ecotypes and cultivated races. The non-taxonomic literature is plagued by a plethora of technical names for the variants of *C. sativa*. Furthermore, in agricultural, horticultural, chemical and pharmacological publications, it is not uncommon to find Latin binomials that have no validity, since they were never published in accordance with the internationally recognized rules of botanical nomenclature.

As early as 1869, De Candolle recognized what he considered to be true botanical varieties of *C. sativa*, offering detailed descriptions of them: α *kif*, β *vulgaris*, γ *pedemontana*, δ *chinensis*.[189] Botanists do not now accept true varieties within *C. sativa* because they cannot define them, and even the agricultural and pharmacological specialists who sometimes treat them as though they were true varieties admit that they are not stable.[189,237] It must be recognized that this problem has arisen primarily because of a confusion of concepts: the true botanical *varietas* is genetically distinct, whilst the polymorphism rampant in *C. sativa* may be non-genetic, giving rise to variations that might better be called races, ecotypes, cultivars, chemovars or other appropriate terms.[189]

Although most modern botanists have held that *Cannabis* is monotypic, there has been opinion to the contrary for many years. Lamarck described *C. indica* in 1783 as a species of "India," distinguishing it from *C. sativa* in growth habit and morphological characteristics and implying, by detailing its strong narcotic properties, that it was more potent than *C. sativa*.

Through the years, most taxonomists have tended not to recognize *C. indica* as distinct, but the binomial—or the alternate *C.*

sativa var. *indica*—has persisted in the chemical and pharmacological literature.

In 1924, Janischewsky described *C. ruderalis* as a species differing from *C. sativa* primarily in morphology of the achene and
size of stem and leaves and ranging from northern European
Russia into western Siberia and central Asia. Other Russian botanists who have studied *Cannabis* in the field maintain that the
genus comprises several species.

Notwithstanding the great economic importance of *Cannabis*
and its long association with man and agriculture, little taxonomic
work has been carried out on the genus. From the period of Linnaeus and Lamarck to the recent Russian studies, no taxonomic
botanists have focussed research specifically on the genus *Cannabis*. Schultes and his colleagues, who have very recently initi-
ated taxonomic and cytologic investigations, now believe that the
genus comprises three species: *C. sativa, C. indica,* and *C. ruderalis.*

What differences, if any, exist in the chemical composition of
the several species it is not yet possible to state. In addition to
the confusion characteristic of *Cannabis* nomenclature, the problem has been complicated by failure of chemists to have voucher
specimens identified and filed away in herbaria. Even more uncertainty was engendered by the recognition of the great chemical variation in races of *C. sativa* or in individual races of this
species grown under differing conditions. Most of the chemical
studies reported in the literature were based undoubtedly on *C.
sativa.* Some chemists may have had true *C. indica* at hand.
Probably relatively few analyses were done on material attribut-
able to *C. ruderalis.*

Since *C. sativa* is a triple-purpose economic plant, it has, over
its thousands of years as an important economic plant, been
selected for characteristics desired by the peoples of the area
where it was cultivated. Where the narcotic properties led to its
role in religious rites, races rich in the intoxicating compounds
tended to be selected; where the nutritive value of the seed-oil
was important, races high in this constituent were selected; where
the plant has been valued for its fibre, races productive in
long and strong fibre were those most desired. It is still not un-

Human experiments with ground nutmeg depleted of its volatiles have failed to show psychopharmacological responses.[403] The volatile fraction consists mainly of two groups of compounds: the terpenes and the aromatic ethers. Although the terpene hydrocarbons constitute by far the larger portion of the volatile fraction, they are usually considered to be of biological effectiveness principally as irritants. The aromatic ethers, then, would seem to be the most likely source of psychotropic activity of nutmeg. Table II shows the structures of the compounds found in the aromatic fraction.[359]

Of the primary constituents, myristicine is by far the most abundant and, for this reason, was tested specifically for psychotropic activity. Doses of 400 mg of myristicine, almost twice the amount present in 20 gm of nutmeg (20 gm being assumed to be the quantity required to produce psychotropic effects) were given to human volunteers and the observed symptoms were at least suggestive of the psychotropic effects in six out of ten subjects.[403] Other aromatic compounds such as elemicin and safrole may have some synergistic activity which would be necessary to produce the full nutmeg syndrome. In pharmacological tests, a myristicine-elemicin fraction of oil of nutmeg produced many of the characteristics of crude ground nutmeg, but again it lacked adequate potency to explain the nutmeg intoxication syndrome on a quantitative basis. Nutmeg and the synthetically produced myristicine demonstrate a mild degree of monoamine oxidase inhibiting activity by in vitro and in vivo tests.[402]

Myristica fragrans Houttuyn, Handleid. 2 (1774) 333.

Tree up to 50 (rarely 60) feet in height, spreading, dioecious (occasionally monoecious), with superficial roots; bark grey. Leaves alternate, estipulate, short-petiolate, prominently pinnatinerved (8–11 pairs), glabrous, elliptic or oblanceolate, acuminate, basally acute, dark green, shiny above, much paler beneath, 5–15 cm long, 2–8 cm wide, aromatic. Staminate and pistillate inflorescences similar, few-flowered (1–10 in former, 1–3 in latter), axillary umbellate cymes, peduncle up to 1.5 cm long. Flowers usually unisexual, very rarely hermaphroditic, fragrant, yellow, waxy, fleshy, glabrous, apetalous; calyx campanulate, basally nectiferous, 3 lobes triangular, acute, reflexed: staminate flowers 6–7 mm long, stamens 8–12, anthers adnate to central column and laterally connate to each other; pistillate flowers up to 1 cm long, ovary sessile, puberulent, 1-celled, stigma short, bifid. Fruits drupaceous, pyriform or occasionally subglobose, nodding, yellowish, 6–10 cm long, longitudinally with circumferential groove along which yellow pericarp splits into 2 valves at maturity. Seeds dark brown-purple, shiny, ovoid, 2–3 cm long, enclosed in bright red or orange-red laciniate aril forming close network around seed.

Myristica fragrans is thought to be native of eastern Malaysia. This species has not been found in an undoubtedly wild state, but the two other species placed with *M. fragrans* in Warburg's series *Fragrans*—obviously close allies—are native and occur wild in Halmahera and New Guinea.

There are about six "races" of *M. fragrans*, differing in minor characters such as the shape, size and aroma of the nuts. Commercial cultivation today centres mainly in the Spice Islands or Moluccas, in Penang and other Malaysian Islands and in Grenada in the West Indies.

Virola Aublet

A genus of 45 to possibly 60 species of tropical forest trees of Central and South America, especially abundant in the Amazon Valley.

A number of Indian tribes of the northwest Amazon (in Brazil and Colombia) and the uppermost Orinoco (in Venezuela) employ hedonistically and ceremonially a highly hallucinogenic snuff made from the blood-red bark resin of several species of jungle trees of the genus *Virola*.

The snuff has various names, according to the tribe or locality, but the most commonly recognized terms for it are *yakee* and *yato* in Colombia, *paricá*, *epená* and *nyakwana* in Brazil.[341,350]

In 1909, Koch-Grünberg, the German anthropologist, referred to a snuff prepared from a tree-bark by the Yekwana Indians in the headwaters of the Orinoco. The witch doctor inhaled the snuff, known as *hakudufha*, during ritualistic cures.

This is a magical snuff, exclusively used by witch doctors and prepared from the bark of a certain tree which, when pounded up, is boiled in a small earthenware pot, until all the water has evaporated and a sediment remains at the bottom of the pot. This sediment is toasted in the pot over a slight fire and is then finely powdered with the blade of a knife. Then the sorcerer blows a little of the powder through a reed . . . into the air. Next, he snuffs, whilst, with the same reed, he absorbs the powder into each nostril successively. The *hakudufha* obviously has a strongly stimulating effect, for immediately the witch doctor begins singing and yelling wildly, all the while pitching the upper part of his body backwards and forwards.[202]

The first definitive association of a snuff with this myristicaceous genus appeared in 1938, when the Brazilian botanist Ducke wrote that the "Indians of the upper Rio Negro use the dried leaves of this species [*V. theiodora*] and of *V. cuspidata* in making a snuff powder that they call *paricá*." [85] In 1939, in discussing the leguminous tree *Anadenanthera peregrina*, the seeds of which provide an hallucinogenic snuff in the Orinoco basin, he asserted: "Martius and other writers attribute to this species the source of the narcotic *paricá* employed by certain Amazonian Indians. . . . Notwithstanding, according to information which I obtained from the natives themselves in two localities in the upper Rio Negro, the paricá-powder comes from the leaves of species of *Virola*. . . ." [86] Although it is now certain that the leaves are not utilized in snuff-making, Ducke's report represents apparently the earliest identification of this narcotic with *Virola*.

Virola-snuff was first described in detail and identified as to species with voucher specimens in 1954, as the result of ethnobotanical studies in Amazonian Colombia. Indians in the area of the Vaupés were preparing a brownish snuff, known amongst the Puinave as *yakee* and the Kuripako as *yato*. Highly narcotic, it was taken only by witch doctors for the diagnosis and treatment of disease, for prophecy and divination and for purposes of magic.[327]

The Indians resident in the Vaupés—Barasana, Makuna, Puinave, Kabuyarí, Kuripako and others—strip the bark from jungle trees early in the morning, before the sun heats up the trunk, and scrape off the soft inner bark with its reddish resinous exudation. The scrapings are kneaded in water which, when strained, is boiled down to a thick syrup. When this syrup has sun-dried, it is pulverized, sifted and mixed with the ashes of the bark of a wild species of *Theobroma*—*T. subincanum*. The resulting snuff is powerful, causing an intoxication which, it is reported, has occasionally led to the death of witch doctors.

The natives in the Colombian Vaupés employ two, and possibly three, species of *Virola* for the preparation of the snuff: *V. calophylla, V. calophylloidea* and possibly *V. elongata.*[327]

Gradually, studies indicated that perhaps the most intensive use of *Virola*-snuffs might be found to the east, centering amongst the several related Indian groups inhabiting the headwaters of the Orinoco in Venezuela and the Brazilian territory north of the Rio Negro. These Indian groups are variously known to anthropologists as the Kirishaná, Shiriana, Karauetari, Karimé, Parahuri, Surará, Pakidái, Yanomami and other terms, but the general term Guaiká or Waiká has been applied to them. They usually refer to their snuff as *epená* or *ebene* or *nyakwana.*[356,453]

Unlike the Indians to the west, these groups employ the snuff occasionally in daily life and quite individually as well as ceremonially; its use is not restricted to the witch doctors but is the prerogative of all male members of the tribe above the age of 13 or 14 years. Furthermore, the snuff is absorbed in excessive—even frightening—amounts, and it appears to be a stronger preparation than that prepared amongst the Colombian Indians.

The principal, and possibly the only, species utilized by the Waiká is *V. theiodora*. Other species have been indicated in the literature as the source of the snuff in the Rio Negro: *V. calophylloidea, V. cuspidata, V. punctata* and *V. rufula.*[39] Since these species give positive alkaloid reactions to spot tests in the field, they may be the source of an intoxicant, but all specimens submitted or illustrated in the literature have been referable to *V. theiodora*.

Several different methods of preparing epená or nyakwana characterize the Waiká. Sometimes, the soft inner layer of the bark is scraped, and the dried shavings are gently roasted over a slow fire. They are then stored until needed for preparation of a new batch of snuff, when they are crushed and pulverized, trit-

Figure 15

VIROLA
calophylloidea
Markgraf

Figure 16

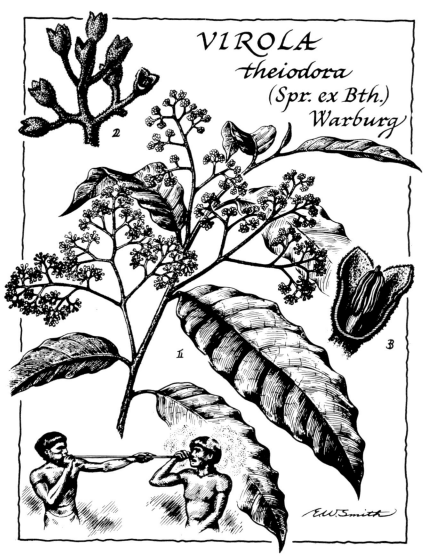

Figure 17

urated in a mortar and pestle of a fruit of *Bertholletia excelsa.* The powder is then sifted to a very fine, homogeneous chocolate-brown, highly pungent dust. Next, a powder of the dried leaves of an aromatic weedy plant, *Justicia pectoralis* var. *stenophylla,* is prepared and added to the brown dust of *Virola*-resin in approximately equal amounts. A third ingredient is the ash of the bark of a rare leguminous tree, *Elizabetha princeps,* known by the Waiká as *amá* or *amasita.* The hard grey outer bark is chopped into small pieces and set in a glowing fire, then removed and allowed to reduce itself to ashes. When the ashes are added in equal amounts to the mixture of the *Virola-Justicia* powder, the resulting snuff, ready for use, is rather greyish and extremely fine.[350]

Other Waiká Indians, who make snuff only occasionally for ceremonial purposes, follow a different procedure. The bark of *V. theiodora* is stripped in the forest from freshly felled trees. A fire is built at the site of the felling. The strips of bark, about two feet long and six inches wide, are laid on or near the fire and gently heated to cause a copious "bleeding" of the red resin which is gathered in an earthenware pot. The resin itself is then boiled down to a thick consistency which, upon cooling, crystallizes into a beautiful amber-red resin. This is then meticulously ground up and reduced to an extremely fine powder which, without any admixture, is nyakwana snuff. Powdered leaves of *Justicia* may occasionally be added "to make the snuff smell better," but the snuff made from the *Virola*-resin alone is highly toxic, and there is, consequently, no real need for the admixture of the *Justicia* powder.[278,279,350]

The effects of *Virola* intoxication vary, but amongst the Indians, they usually include initial excitability, setting in within several minutes from the first snuffing. Then follows numbness of the limbs, twitching of the facial muscles, inability to coordinate muscular activity, nausea, visual hallucinations and, finally, a deep, disturbed sleep. Macropsia is frequent and enters into Waiká beliefs about the spirits that dwell in the plant.[29,327,350]

A still unresolved aspect of *Virola* toxicology is its employment by the Waiká as an arrow poison. The resin is applied directly to arrows and darts, taken fresh from the bark, and is merely heated slightly and smoked to solidify each of the many coats that are spread on the point. The chemical explanation of this activity is still far from understood.[350]

An interesting way of using *Virola* resin by oral administration has recently been discovered in Amazonian Colombia. The Witoto, Bora and Muinane Indians living on northern affluents of the Rio Putumayo ingest small pellets of the resin of *V. theiodora* to see and communicate with "the little people." The *Virola* tree is known by the Witoto as *oo-koó-na, koo-troo-koo* by the Muinane.[344]

These Indians rasp the inner part of the freshly stripped bark, roll the rasped tissue into balls and express the resin from it into a pot of water which is boiled for five or six hours, until the mass becomes a thick syrup that sticks to the wooden paddle that has been used to stir the boiling mixture.

While this operation is progressing with the resin of *Virola*, another Indian reduces to ashes the bark of *Gustavia poeppigiana*. The Witoto call this tree *hĕ-rog*. The ashes are put into a funnel made of leaves, and cool water is poured over the ashes and allowed to seep until no more cloudiness is seen in the filtrate. The water is then boiled down slowly until a greyish residue or "salt," called *lĕ-sa* in Witoto, is left.

The thickened resin is then rolled with the fingers into tiny pellets the size of coffee beans, and these are rubbed in the salt-like residue from the leached-out bark ashes. The pellets, thus coated, are ingested whole or dissolved in water and drunk. From three to six pellets are taken initially, and the intoxication is said to begin within five minutes and last up to two hours. More pellets may be taken when the effects of the drug begin to lessen in intensity.[344]

One group of Makú Indians in Amazonian Colombia drink the resin of *V. elongata* with no preparation for its intoxicating effects.[360]

There are vague suggestions that witch doctors in Venezuela may smoke the bark of *V. sebifera*.[348] These suggestions are found on the labels of herbarium specimens from Venezuela in the Field Museum. ". . . the inner bark is dried and smoked by witch doctors for smoking at dances when curing fevers; it is

Figure 18. Waiká Indian gathering resin from *Virola theiodora* for preparation of hallucinogenic resin nyakwana. Rio Tototobí, Amazonian Brazil.
Photograph by R. E. Schultes.

very strong." Another collection of the same species from the
upper Orinoco by the same botanist, bears the information: "Indians boil bark and use to drive away evil spirit." The Indian
names of this species are given as *wircaweiyek* or *orika-bai-yek*
and *piassám*. In view of the discovery of alkaloids in the bark of
V. sebifera,[69A] may these reports not indicate that perhaps the
natives employ it likewise, possibly by smoking, as an hallucinogen?

The biodynamic properties of *Virola*-resin were first thought to
be due to myristicine. Recent studies, however, have established
the presence of several interesting tryptamines in some species
of *Virola* used by aborigines in the preparation of snuff powders.[11]
The following tryptamine derivatives and β-carbolines have been
found:

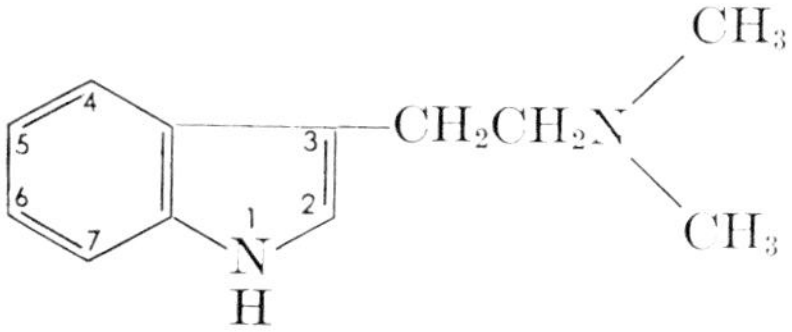

N,N-Dimethyltryptamine
(DMT)

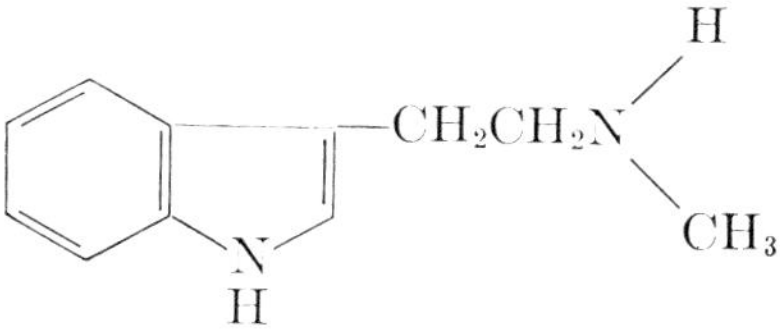

N-Methyltryptamine
(MMT)

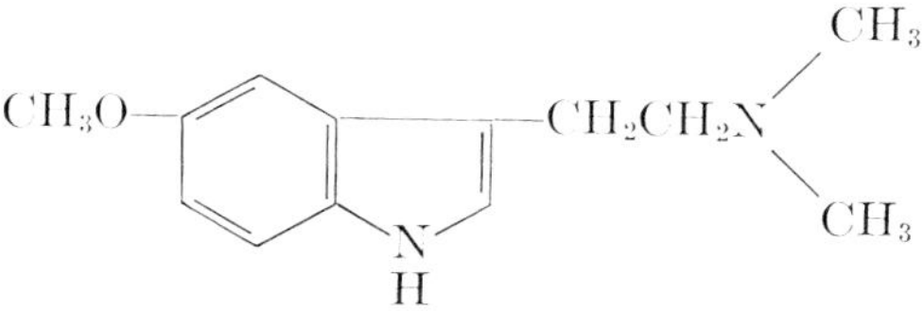

5-Methoxy-N,N-dimethyl-
tryptamine   (5-MeO-DMT)

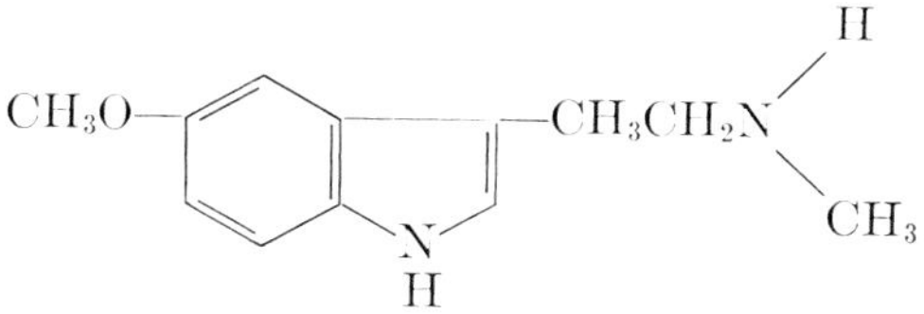

5-Methoxy-N-methyl-
tryptamine (5-MeO-MMT)

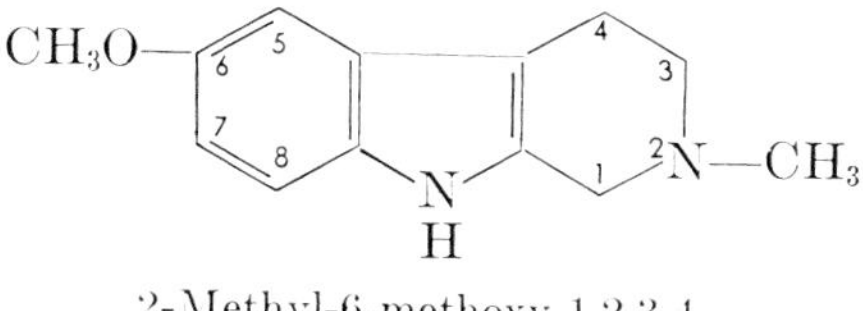

2-Methyl-6-methoxy-1,2,3,4-
tetrahydro-β-carboline
(6-MeO-THC)

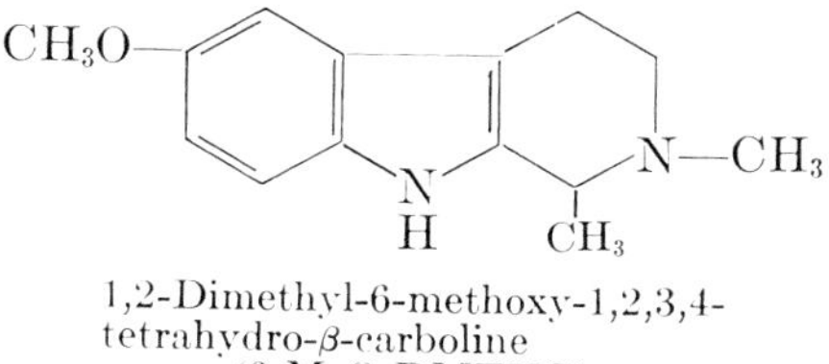

1,2-Dimethyl-6-methoxy-1,2,3,4-
tetrahydro-β-carboline
(6-MeO-DMTHC)

TABLE III

OCCURRENCE OF ALKALOIDS IN THREE SPECIES OF *VIROLA*.

Species	Part of plant	mg/100gm dry plant	Alkaloids	Percent
V. theiodora Origin: Manáus, Brazil	Bark	250	DMT	52
			5-MeO-DMT	43
			6-MeO-THC	4
			MMT	1
	Root	17	5-MeO-DMT	62
			DMT	22
			5-MeO-MMT	15
	Flow. shoots	470	DMT	93
			MMT	7
	Leaves	44	DMT	99
			5-MeO-DMT	
V. theiodora Origin: Tototobí Brazil	Bark	65	5-MeO-DMT	95
			DMT	5
	Leaves	21	DMT	98
			MTHC	2
V. calophylla Origin: Manáus Brazil	Bark	9	DMT	91
			5-MeO-DMT	9
	Root	1	DMT	87
			5-MeO-DMT	13
	Flow. shoots	193	DMT	96
			MMT	4
	Leaves	155	DMT	96
			MMT	4
V. rufula Origin: Manáus Brazil	Bark	200	5-MeO-DMT	95
			DMT	4
			5-MeO-MMT	
			6-MeO-THC	
	Root	144	5-MeO-DMT	94
			5-MeO-MMT	4
			DMT	1
			6-MeO-THC	
	Leaves	98	DMT	94
			MMT	6

One Indian snuff—that of the Waiká—prepared solely from resin of V. *theiodora*, proved to be unusually high in alkaloid content (11%).[178] It consisted mainly of 5-MeO-DMT, with lesser amounts of DMT.[10,180]

There is appreciable variation in alkaloid concentration in different parts (leaves, bark, root) of the same plant, as can be seen in Table III.

Tryptamine derivatives are easily synthesized. DMT was thus produced for the first time in 1931 by Manske.[230] Later, a number of improved syntheses for DMT and methoxylated tryptamine derivatives were described.[104,380,387,416]

The constituent 2-methyl-6-methoxy-tetrahydro-β-carboline

Figure 19. *Virola theiodora:* flowering branches. Manáos, Brazil. Photograph by R. E. Schultes.

was synthesized from N-methyl-5-methoxy-tryptamine and form-aldehyde, whereas 1,2-dimethyl-6-methoxy-tetrahydro-β-car-boline was prepared by an analogous reaction from N-methyl-5-methoxy-tryptamine and acetaldehyde.[11]

The psychotomimetic properties of synthetic N,N-dimethyl-tryptamine were tested by Szàra [44,309,390] and later confirmed by other authors.[20,406] The effective dose for man was found to be approximately 1 mg/kg intramuscularly. The hallucinogenic effects set in rapidly and lessen in intensity and disappear after 50 to 60 minutes.[391] DMT is not active orally. 5-Methoxy-N,N-dimethyl-tryptamine (5-Me-DMT) has been found to be more active in a conditioned avoidance response test than 5-hydroxy-N,N-di-methyltryptamine (bufotenin) or DMT.[115] Benington and others report that the effect of 5-MeO-DMT on cat behaviour is dramatic.[34]

No data concerning the hallucinogenic activity of 6-MeO-THC and of 6-MeO-DMTHC are available. These β-carbolines may, however, act as monoamine oxidase inhibitors.

It may be that the use of the plant materials in the form of a snuff has some effect in modifying the hallucinogenic activity of the tryptamine and β-carboline derivatives.

Virola theiodora (*Spr. ex Benth.*) *Warburg* Nova Acta Acad. Leop.-Carol. 68 (1897) 187.

Slender tree, 25–75 feet in height, trunk cylindrical up to 1½ feet in diameter, bark smooth, brown mottled with grey patches. Branchlets slightly red-brown tomentellous, becoming glabrous. Leaves (with pronounced fragrance of tea when dried) firm-papyraceous, sometimes even thick-chartaceous, often sparsely glandular-punctate, oblong to broadly ovate, basally obtuse to cordate, apically long-acuminate, marginally usually sinuate, 9–35 cm long, 4–12 cm wide, upper surface glabrous, dark green, nitid, nether surface sparsely stellate-puberulent, secondary veins 9–20, usually very prominent, ascending and arcuate, petiole subterete, mostly 4–15 cm long, often brown-tomentellous. Staminate inflorescences many-flowered, paniculate, usually shorter than leaves, up to about 15 cm long (usually shorter), often brown- or golden-brown-tomentellous, usually becoming glabrous, bracts 2.5 cm long, deciduous. Pistillate inflorescences shorter. Staminate flowers strongly pungent, single or in clusters of 2–10, pedicel about 2 mm long, perianth thin, puberulent within and without, infundibuliform, 1.5–2.5 mm long, sub-acutely lobed about one fourth its length; androecium 2 mm long, filament column thick, 0.5–0.8 mm long, anthers 3–5 (usually 3), 1–1.7 mm long, usually connate, apiculate. Fruit about 5–8 per inflorescence (often fewer), subglobose, 10–20 mm long, 8–15 mm in diameter, usually slightly apiculate, glabrescent when mature, pedicels 3–4.5 mm long, aril membranaceous, laciniate about one half its length.

Distributed mainly in the western Amazonia of Brazil and Colombia, possibly also in adjacent parts of Peru and Venezuela; especially abundant in the Rio Negro basin. A tree of well drained forests.

Other species used include the following:

Virola calophylla *Warburg*, Nova Acta Leop.-Carol. 68 (1897) 231.

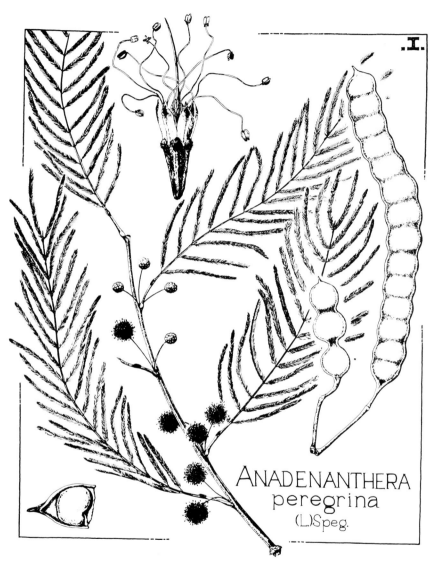

Figure 22. Drawn by I. Brady.

grey to black, lenticellate, often armed with stout, conic spines or cuneate projections. Branches unarmed, branchlets puberulent. Leaves bipinnate, 12–30 cm long (including petiole), puberulent, glabrescent in age, main rhachis sulcate, pinnae 10–30 or more, each 2–5 cm long or longer, opposite or subopposite. Petiole with oval gland, 0.5–5 mm long near base of leaf, rhachis occasionally with 7–10 smaller glands. Leaflets 25–80 pairs, membranaceous, linear, oblong or lanceolate, straight or falcate, basally oblique or truncate, apically apiculate, 2–8 mm long, 0.5–1.5 mm wide, upper surface dark green, nitid, nether surface paler. Inflorescence normally compact, globose-capitate head of 35–50 very small white flowers in terminal or axillary racemes, 10–18 mm in diameter, in fascicles of 1–5; peduncles 1.7–4 mm long, puberulent. Flowers membranaceous, calyx campanulate, 5-lobed, 0.5–2.5 mm long, petals 5, free or slightly connate, 2–3.5 mm long. Stamens 10, free, 5–8 mm long, anthers eglandular or with minute stipitate gland in bud. Pod leathery or subligneous, brownish, dull, scurfy to verrucose, broadly linear or strapshaped, 5–35 cm long (exclusive of peduncle), 1–3 cm wide, usually contracted between seeds. Seeds 3–15 or more (usually only a few ripening in pod), very thin, flat, orbicular to suborbicular, 10–20 mm in diameter, brown to black, nitid.

Anadenanthera peregrina occurs in northern South America (northern Brazil, British Guiana, Colombia, Venezuela) and (possibly naturalized) in the West Indies, growing primarily in open plains areas, scrub or wastelands, savannahs along watercourses, woody hillsides, on open ridges, preferring clay or sandstone soils and, at least in South America, semiarid habitats.

Most of the literature refers to this species under its older technical name, *Piptadenia peregrina (L.) Bentham,* in Journ. Bot. 4 (1841) 340.

A more southern variety—*A. peregrina* var. *falcata*—has been described. Known from southern Brazil and Paraguay, it differs morphologically in having more numerous pinnae but fewer, shorter and straighter leaflets, yellow flowers and shorter, falcate (not straight) pods.

Anadenanthera colubrina occurs in eastern Brazil, and its variety *A. colubrina* var. *cebil* is known from Argentina, Bolivia, Paraguay, Peru and several localities in southeastern Brazil.

The nomenclatural history of *Anadenanthera* is extremely complex. Altschul's recent monographic study has gone far towards ending long standing confusion concerning both generic and infrageneric categories.[15]

> *Anadenanthera peregrina* (*L.*) *Spegazzini* var. *falcata* (*Benth.*) *Altschul* Contrib. Gray Herb. Harvard Univ. 193 (1964) 50.
>
> *Anadenanthera colubrina* (*Vell.*) *Brenan*, Kew Bull. 2 (1955) 182.
>
> *Anadenanthera colubrina* (*Vell.*) *Brenan* var. *cebil* (*Griseb.*) *Altschul* Contrib. Gray Herb. Harvard Univ. 193 (1964) 53.

Cytisus Linnaeus

There are about 80 species of *Cytisus* known from the Atlantic islands, Europe and the Mediterranean area. Some of the species are valued as ornamentals; a number of them are toxic.

A recent report indicates that Yaqui medicine men from northern Mexico employ *C. canariensis* (also known as *Genista canariensis*)—the genista of florists—as an hallucinogen in their practices.[97] This is an introduced Old World species, native to the Canary Islands, and there is apparently no record of its use as a narcotic elsewhere.

The genus *Cytisus* is rich in cytisine,[444] an alkaloid of the lupine group, for which hallucinogenic activity has not been pharmacologically demonstrated (see under *Sophora*). Further chemical study is indicated to determine whether or not there are additional constituents in *C. canariensis* which might explain its native use.

> *Cytisus canariensis* (*L*). *O. Kuntze*, Rev. gen. Pl. 1 (1891) 177

Much-branched, evergreen, unarmed shrub up to 6 feet in height. Branches villous-pubescent; branchlets grooved. Leaves trifoliate, petiolate, persistent; leaflets cuneate, obovate to oblong-obovate, obtuse or rounded, 0.8–1.5 cm long, pubescent on both surfaces. Inflorescences dense, short terminal racemes, many-flowered. Flowers fragrant, bright yellow, 15 mm long; calyx campanulate, 2-lipped, teeth short, as long as or only slightly longer than wide, stamens monadelphous, style curved. Pods flat, dehiscent, pubescent, 15–20 mm long. Seeds with basal callous appendage.

Native of the Canary Islands but widely cultivated in milder climates.

Mimosa Linnaeus

The genus *Mimosa*, belonging to the subfamily Mimosoideae, and closely allied to *Acacia*, comprises about 500 tropical and subtropical species of herbs and small shrubs, mostly of American distribution but some native to Africa and Asia.

Several tribes of eastern Brazil employed the root of the shrub *M. hostilis* in the preparation of a "miraculous drink" known locally as *ajuca* or *vinho de jurema*. The tribes known to use it were the Kariri, Pankaruru, Tusha and Fulnio of the states of Pernambuco and Paraíba. The drink was consumed ceremonially in the ajuca ritual.[223-225]

The roots of *M. hostilis*, which grows in the dry, scrubby ca-atinga vegetation of the area, are the source of the intoxicant. This jurema cult is apparently ancient, having formerly been practiced by a number of tribes, mostly now extinct or wholly acculturated: Guegue, Acroa, Pimenteira, Atanayé. An early report of the jurema ceremony dates from 1788. Another, dating from 1843, asserted that, in a number of tribes, jurema was taken to "pass the night navigating through the depths of slumber" and, by relating it to the use of *paricá* (presumably a species of the related genus (*Anadenanthera*) and *ipadú* (*Erythroxylon coca*), seems to suggest hedonistic employment of jurema.[120]

This potent hallucinogenic drink merits deeper study. There is a question as to whether or not living groups of Indians still employ it, but, amongst those natives who utilized it until recently, groups of priests, warriors or strong young men and old women singers participated in the ceremony, all kneeling with heads bowed to receive their portion of the drink. At one time, the jurema ceremony was performed especially just prior to sallying forth to war. A recent description of the jurema cult records that

> an old master of ceremonies, wielding a dance rattle decorated with a feather mosaic, would serve a bowlful of the infusion made from jurema roots to all celebrants, who would then see glorious visions of the spirit land, with flowers and birds. They might catch a glimpse of the clashing rocks that destroy souls of the dead journeying to their goal or see the Thunderbird shooting lightning from a huge tuft on his head and producing claps of thunder by running about.[223-225]

Apparently, several species of *Mimosa* are generically called jurema in northeastern Brazil. One of the several kinds, *jurema prêta*, is *M. hostilis*, from which the intoxicant is prepared. It is said sometimes to be called also *jurema branca*, although this name may refer as well to *M. verrucosa*, from the bark of which a stupefacient is said to be derived.[241]

An alkaloid isolated from the roots of *M. hostilis* in 1946 was named nigerine.[120] In more recent studies, a single alkaloid N,N-dimethyltryptamine was extracted in 0.57 percent yield from the roots of *M. hostilis*, and the identity of nigerine with this alkaloid was established.[256] N,N-dimethyltryptamine is also one of the main constituents of a related hallucinogenic genus: *Anadenanthera*.

Mimosa hostilis (*Mart.*) *Bentham*, Trans. Linn. Soc. 30 (1875) 415.

Bushy treelet, branchlets minutely viscid-puberulent, stipules subulate, spines sparse, strong, straight, basally swollen, 5–6 mm long. Leaves bipinnate, 3–4 cm long, 4–6-jugate. Pinnae multi-jugate (8–18), 2.5–3 cm long. Leaflets rather thick, obliquely oblong, very obtuse, 1–2 mm long, aculeate, puberulent. Inflorescence loosely cylindrical, 5.5–6 cm long, rhachis puberulent, bracts small, spathulate. Flowers white, 4-merous, 8-androus, calyx minute, obtusely 4-lobate, corolla deeply 4-fid, stamens twice as long as corolla or longer. Pod sessile or short-stipitate, 2.5–3 cm long, about 3 mm in diameter, viscid-puberulent, valves membranaceous, breaking into 4–6 sections.

Known from eastern Brazil: Minas Gerais, Bahía, Pernambuco.

MIMOSA hostilis (Mart.) Benth.

Figure 23. Drawn by J. B. Clark.

Sophora **Linnaeus**

Sophora belongs to the Sophoreae of the subfamily Papilionoideae. Related to *Myroxylon, Baphia* and the highly toxic genus
Ormosia, Sophora comprises approximately 50 species ranging
through tropical and warm temperate parts of both hemispheres.

SOPHORA secundiflora (Ort.) Lag.

Figure 24. Drawn by J. B. Clark.

The beautiful scarlet seeds of S. *secundiflora*, known as *mescal beans*, *red beans*, *coral beans* or, in Mexico, as *frijolillos* or *colorines*, once formed the basis of a vision-seeking cult amongst the Indians of northern Mexico, Texas and New Mexico.[206,319,331,345]

Figure 25. *Sophora secundiflora*. Uvalde County, Texas. Photograph by D. S. Correll.

An early Spanish explorer of the coast of Texas, Cabeza de Vaca, mentioned mescal beans as one of the articles of trade amongst the natives in 1539. According to the Stephen Long expedition, the Arapaho and Iowa tribes in 1820 were using the large red beans as a medicine and narcotic. Mescal beans have been found in archaeological sites, all dated before 1000 A.D., sometimes with evidence of their possible ritualistic utilization. They have been recorded from a number of archaeological sites in caves and rock shelters in southwestern Texas. Material from sites in northern Mexico has been carbon-dated to between 1500 B.C. to 200 A.D., thus substantiating the antiquity of the use of this toxic seed. Although, as Campbell has written, "the presence of mescal beans in . . . sites, even when included in containers holding utilitarian as well as non-utilitarian objects, does not necessarily signify the presence of a mescal bean cult. . . . There is additional archaeological evidence which does suggest the presence of a prehistoric cult that may have involved the use of the mescal bean." [59]

At any rate, there has existed in modern times a well developed mescal bean cult amongst a number of North American Indians: the Apache, Comanche, Delaware, Iowa, Kansa, Omaha, Oto, Osage, Pawnee, Ponca, Tonkawa and Wichita tribes. Still other tribes esteemed the bean as a medicine or fetish but failed, apparently, to develop a cult surrounding its use. In the cult—known variously as the Wichita Dance, Deer Dance, Whistle Dance, Red Bean Dance and Red Bean Society—the seeds were known to have been employed ritualistically or as an oracular or divinatory medium for inducing visions in initiatory rites and as a ceremonial emetic and stimulant.[59,181,207,209,400]

There are many similarities and parallels between certain aspects of the modern Peyote Cult and the Red Bean Dance, and both obviously had a southern origin because of the natural distribution of the plants involved. An ethnobotanically and pharmacologically most interesting practice is the reported mixing in a narcotic drink of peyote and mescal beans amongst the Comanche, Oto and Tonkawa. This drink, employed probably in transitional periods between the dying out of the Red Bean Dance and the establishment of the Peyote Cult in the last century, must indeed have been a potent, if not a dangerous narcotic preparation. It may possibly be responsible for the confusion in certain early literature of the terms *mescal beans* and *peyote*.[59,181,206]

It appears probable that the Red Bean Dance was pre-peyote in its apogee amongst the Plains tribes, where its role as a sacred narcotic was lost or forgotten with the arrival of the much safer hallucinogenic cactus. Even today, Kiowa, Comanche and other leaders or "roadmen" of the peyote ceremony often wear, as part of their ornamental dress, a necklace of beans of S. *secundiflora*.

Chemical investigation of *Sophora secundiflora* dates back to the end of the last century. An alkaloid, sophorine, was isolated,[272] and it later proved to be identical with cytisine, which is found in many genera in the Leguminosae.[213] The chemical structure and absolute configuration of cytisine, which belongs to the group of the lupine alkaloids, have been completely elucidated.[249]

Cytisine belongs pharmacologically to the same group as nicotine.[452] It is a powerful poison causing nausea, convulsions and

Figure 26. Bead necklace of *Sophora secundiflora* seeds used by leader of peyote ceremony. Kiowa tribe, Anadarko, Oklahoma. Courtesy of Botanical Museum, Harvard University.

death through failure of the respiration. No hallucinogenic activity has been pharmacologically reported.

Chemical analysis of *S. secundiflora* using modern methods should be repeated in order to determine whether or not other alkaloids occur in this plant which could explain its native use as an hallucinogen.

Sophora secundiflora (*Ort.*) *Lagasca ex De Candolle,*
Cat. Hort. Monsp. (1813) 148.

Shrub or small tree 25–40 feet in height, trunk 6–8 inches in
diameter dividing near base into upright branches forming narrow crown. Branchlets tomentose, glabrescent or nearly so.
Leaves evergreen, persistent, 10–15 cm long, petiolate, petioles
puberulent, basally slightly swollen. Leaflets 7–11, subsessile,
chartaceous, oblong-elliptic, 1.5–6 cm long, apically obtuse or
sometimes mucronate, basally broadly cuneate, upper surface
deep yellow-green, nitid, nether surface sericeous, especially
when young. Flowers in terminal, 1-sided racemes 5–10 cm long,
sweet-scented, violet-blue (standard with few darker spots at
base), 2–3 cm long, calyx campanulate, petals short-unguiculate,
ovary white-sericeous. Pod hard, woody, terete, strongly constricted between seeds, 2.5–20 cm long, 1.7 cm in diameter,
hoary-tomentose, indehiscent, 1–8 seeded. Seeds bright scarlet,
ovoid, about 1.5 cm long, hilum small, paler, testa hard.

Ranging from northern Mexico to Texas and New Mexico,
along streams in thickets and small groves, usually on limestone
soil. It is now often planted as an ornamental.

MALPIGHIACEAE

The Malpighiaceae, belonging to the order Geraniales, is a
tropical family (especially well developed in South America) of
well over 800 species in 60 genera. It is divided into two subfamilies: Planitorae (trees of neotropical distribution) and
Pyramidotorae (lianas and vines of the tropics of both hemispheres). Except for several minor fruit trees, there are no significant economic species, although a number of genera are recognized by native peoples as poisonous or have interesting uses in
folk-medicine. Chemotaxonomic investigation of the Malpighiaceae is long overdue.

Banisteriopsis C. B. Robinson & Small

Banisteriopsis is a tropical American genus numbering about
100 species of vines.

There is a narcotic drink widely employed in northern South

America for prophecy, divination and, in general, as a magic hallucinogen. It has many indigenous names, but it is generally known as *ayahuasca, caapi* or *yajé*. Although not nearly so popularly recognized as peyote or, nowadays, as the sacred mushrooms, this narcotic preparation has nonetheless had an undue share of sensational articles that have played fancifully with the extravagant and unfounded claims concerning the powers of the drink, especially in regard to its "telepathic" properties.

Several species of *Banisteriopsis* are involved in the preparation of this narcotic drink which is made from the bark of the stems in Amazonian Brazil, Bolivia, Colombia, Ecuador, Peru, the Orinoco of Venezuela and the Pacific coast of Colombia and Ecuador. Usually, only one species enters the preparation, but, on occasion, admixtures may be employed.

Two species of *Banisteriopsis* seem to constitute the basic ingredients of the drink: *B. caapi* in the greater part of the geographical range of this narcotic; *B. inebrians* in the very westernmost part of the Amazon basin near the eastern slope of the Andes. These species, like many of the others in this genus, are rather incompletely known, due in part to the paucity of fertile collections available for taxonomic study. Both are jungle lianas that apparently flower very infrequently. It is difficult for the nonbotanist to understand our lack of understanding of specific delimitations of drug plants, the use of which has been known for more than a century.[71,72,242,329]

The earliest report of ayahuasca appears to be that of Villavicencio in his Ecuadorian geography, published in 1858. The source of the drug, he wrote, was a vine employed by the Zaparo, Angatero, Mazán and other tribes of the Rio Napo drainage-area

> to foresee and to answer accurately in difficult cases, be it to reply opportunely to ambassadors from other tribes in a question of war; to decipher plans of the enemy through the medium of this magic drink and take proper steps for attack and defense; to ascertain, when a relative is sick, what sorcerer has put on the hex; to carry out a friendly visit to other tribes; to welcome foreign travellers; or, at last, to make sure of the love of their womenfolk.[414]

Seven years earlier, in 1851, the British plant explorer Spruce had discovered the Tukanoan tribes of the Rio Uaupés in Amazonian Brazil using a liana called *caapi* to induce intoxication, but his observations were not published until later.[382] Unlike Villavicencio, Spruce precisely identified caapi as a new species of the Malpighiaceae. A collection in full flower was taken from the liana from which the drink was prepared, and Spruce drew up a detailed description of it, calling it *Banisteria caapi* from its vernacular name. The description was published by Grisebach. As taxonomic understanding of the family Malpighiaceae grew during the present century, it was shown that this and related species-concepts could not with precision be included in *Banisteria*. In 1931, Morton transferred the concept to the genus *Banisteriopsis*.[242] The correct name now is, accordingly, *Banisteriopsis caapi*. In 1931, Morton described as a related species, likewise employed as an hallucinogen, *B. inebrians* from material from the westernmost Amazon in Colombia.[242]

Spruce was so far ahead of his botanical contemporaries that he collected stems for chemical analysis from the type or original plant in 1851.[343] This material was subjected to analysis in 1969, and after more than a century, gave results comparable with freshly collected plant material.[351] In 1853, Spruce met with the use of caapi amongst the Guahibo Indians of the upper Orinoco basin of Colombia and Venezuela. Here, he reported, the natives "not only drink an infusion, like those of the Uaupés [Vaupés], but also chew the dried stem. . . ." Again, in 1857, in the Ecuadorian Andes, he encountered the Zaparo taking a narcotic known as *ayahuasca*, and he considered it to be "the identical species of the Uaupés, but under a different name." Spruce wrote that although "of the plant itself" Villavicencio "could tell no more than that it was a liana or vine," his "account of its properties" coincided "wonderfully with what I had previously learnt in Brazil."[382]

In the century that followed Spruce's remarkable work, many explorers, botanists and others—von Martius, Orton, Crévaux, Koch-Grünberg, to name only a few—referred to ayahuasca, caapi or yajé, usually casually and often without any botanical

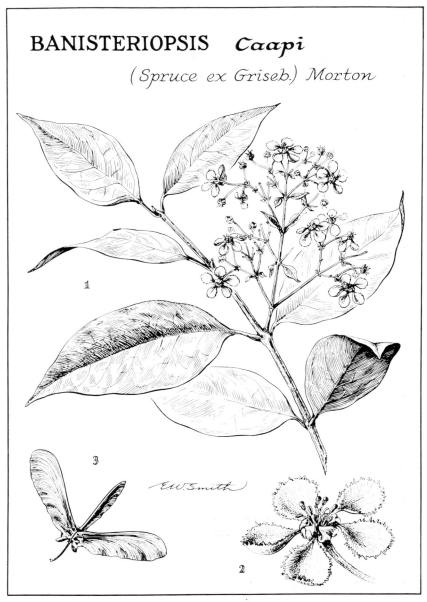

Figure 27

identification beyond the statement that the drug came from a "jungle liana." [246,329,330]

In the years that followed the early work, the area of use of ayahuasca, caapi or yajé was shown to include the Amazon of Peru and Bolivia and even to the rain-forested Pacific coastal regions of Colombia and Ecuador. Several other species of *Banisteriopsis* and even a species of *Mascagnia* were likewise stated to be employed, but the reports are vague and usually without adequate botanical authenticity.[329]

Of outstanding interest was the work in 1922 of Rusby and White in Bolivia and the publication by Morton in 1931 of the detailed field notes made by the meticulous botanical collector Klug in the Colombian Putumayo. Similarly, the studies of the Russian collectors Varanof and Juzepczuk in the Colombian Caquetá in 1925–26 added valuable data. The field work of the Colombian botanist García-Barriga has added appreciably to our knowledge, and an exceptionally complete ethnobotanical study was published in 1965 by Friedberg.[107] Finally, in 1958, Cuatrecasas' monographic study of the Malpighiaceae of Colombia has provided, for the first time, a firm taxonomic basis for clarification of ethnobotanical and phytochemical problems with this most curious of hallucinogens.[69,246,347,349]

Serious complications, however, arose early in attempts correctly to identify ayahuasca, caapi, and yajé. In 1890, a missionary published an article in which he confused the narcotic tree-species of *Datura*, employed amongst the Jívaro, with the malpighiaceous hallucinogen, a confusion that quickly entered pharmacological and chemical literature and which has persisted. The narcotic drink was, at one time, even attributed to a species of *Aristolochia*.[329]

There further arose, mainly from uncritical reading or misinterpretation of Spruce's field notes, a strange presumption that gained very wide acceptance in the literature; it was presumed that, although ayahuasca and caapi were derived from *Banisteriopsis*, yajé was prepared from the apocynaceous *Prestonia amazonica* (*Haemadictyon amazonicum*). Although it has been discredited, this misidentification is encountered throughout the literature in reference to yajé, and occasionally to ayahuasca and

Figure 28. *Banisteriopsis caapi:* in flower and fruit. Grown in Valle del Cauca, Colombia. Photograph by J. Cuatrecasas.

caapi as well. It has confounded the ethnobotanical, chemical and pharmacological study of this drug to an astonishing degree and, unfortunately, still appears in even technical literature.[352]

From a critical evaluation of field work from all sources, it is now amply clear that the two main sources of ayahuasca, caapi and yajé in the Amazon basin, *natéma* in Ecuador and *pinde* along the Pacific coast of Colombia are the barks of *B. caapi* and *B. inebrians.*

The first chemical analyses of *Banisteriopsis* suffered from the lack of reliable botanical determination of the plant material. An alkaloid was isolated by various workers and called "telepathine" [265] or "banisterine." [219] Further investigations by Elger [88] and by Wolfes and Rumpf [447] showed that only one alkaloid was concerned and that it was identical with the already well known harmine, which had been isolated considerably earlier from *Peganum harmala.* In addition to this principal alkaloid, two minor alkaloids, harmaline and d-tetrahydroharmine, were later shown to be present in small amounts.[152]

the Tukanos of the Brazilian part of the Rio Vaupés, but, unfortunately, the plants are still identified only by native names; the admixture said to fortify the drink most strongly—a vine with thickened nodes and known in Tukano as *kuri-kaxpidá*, may possibly represent *Gnetum* sp., a very abundant element of the riverside vegetation.[329,347,349]

Recently, two of the sundry admixtures have attracted attention because phytochemical investigation has substantiated folk uses. More often than not, specialists have explained many of the admixtures in this and other narcotic and poison formulae (curare is a good example) as based on superstition. The two admixtures in question are *B. rusbyana* and *Psychotria viridis*.[347]

It was Poisson who, in 1965, reported that the leaves of *B. rusbyana* contained N,N-dimethyltryptamine in relatively high concentration.[273] This discovery was corroborated by several later investigators.[9,79] Minor components consist of N-methyltryptamine, 5-methoxy-N,N-dimethyltryptamine and bufotenin (formula: see *Virola* section). Furthermore, N-methyltetrahydro-β-carboline was found in trace amounts.[9]

N-methyltetrahydro-β-carboline

The leaves of *B. rusbyana*, known in the Colombian and Ecuadorian Amazon as *oco-yajé*, are one of the important admixtures to the drink prepared from bark of *B. caapi* or *B. inebrians*. The natives add these leaves "to heighten and lengthen" the visual hallucinations of the intoxication. It is now known that there is a chemical basis for the use of this admixture: the resulting drink, containing the β-carboline alkaloids and N,N-dimethyltryptamine, both hallucinogenic indole derivatives, is, in effect, far more potent.

Recent ethnobotanical field studies have shown that, in two distant parts of the Amazon drainage-area, in eastern Ecuador and the Acre of Brazil, *Psychotria* leaves are added to *Bani-*

steriopsis drinks.[278,279,347,349] One of the species is *P. viridis*, employed in both localities; the other has not yet been identified as to species. Chemical analysis has recently indicated the presence in the leaves of *P. nitida* of N,N-dimethyltryptamine, the first time in the Rubiaceae.[78]

In connection with the discovery of N,N-dimethyltryptamine in *B. rusbyana*, it should be pointed out that, in 1957, two chemists reported an analysis of "ayahuasca" and "yajé" from Amazonian Peru. They "identified" ayahuasca as *B. caapi* and yajé as *Prestonia amazonica*, stating that the natives of the Río Napo "commonly consume a mixed extract of the *B. caapi* and *P. amazonica* leaves in the belief that the latter suppress the more unpleasant hallucinations associated with the pure *B. caapi* extracts."[152] The "identification" of the *"Prestonia amazonica"* leaves was made on an aqueous extract of the leaves through the use of the vernacular name *yajé* which, in much of the literature, is so identified. N,N-dimethyltryptamine is unknown in the Apocynaceae, and, in the light of recent studies, it appears possible that the aqueous extract of leaves called *yajé* was, in reality, *B. rusbyana* or *Psychotria viridis*.

Much field and laboratory investigation, preferably interdisciplinary, must be done before a complete understanding of the ayahuasca-caapi-yajé complex is available, notwithstanding the fact that more than a century has elapsed since the first botanical work was done. It is disconcerting that, with the rapid acculturation and civilization of tribe after tribe, the time for pristine investigations of this kind is fast disappearing.

The course of harmine intoxication in intact cats corresponds to the effect of a centrally excitant spasmogen.[30,31] The evident excitant effect of harmine and harmaline is very probably related to their inhibiting effect towards monoamine oxidase (MAO).[270] It is known that this enzyme participates in the decomposition of important biogenic amines, and its inhibition leads to an accumulation of epinephrine and norepinephrine in the organism.

Small doses of harmine (25–75 mg subcutaneously) are reported to produce euphoria in man.[219] Turner and his associates [407] doubted that harmine was psychotomimetically active, although Gershon and Lang [114] found that it caused restlessness and apparent hallucinations in dogs. Pennes and Hoch [260] reported LSD-like effects in mental patients given 150–200 mg intravenously, whereas oral application (300–400 mg) produced, as the only perception disturbance, the impression of a wavelike movement of the environment, as well as parasthesia and a lower sensitivity of the skin to contact and pain stimuli.

In a study with 30 volunteers who were given harmaline hydrochloride orally or intravenously (no indication of the quantity), Naranjo observed withdrawal from the outer world and extreme passivity. The most characteristic effect was the closed-eye contemplation of vivid, bright-coloured imagery. No other psychotomimetic effects such as ecstatic feelings, characteristic for other hallucinogens, were observed.[244]

The hallucinogenic effects of ayahuasca preparations may be the result of the combined activity of harmine and its derivatives with dimethyltryptamine and other tryptamine derivatives which were found in certain admixtures to ayahuasca such as *Banisteriopsis rusbyana* and *Psychotria viridus*. The MAO-inhibitors harmine and harmaline may enhance the effects of the tryptamines. It should, however, be borne in mind that the narcotic is still hallucinogenic even when, as often is the case, it is prepared exclusively from species of *Banisteriopsis* without the tryptamine-rich additives.

Banisteriopsis caapi (*Spr. ex Griseb.*) *Morton*, Journ. Wash. Acad. Sci. 21 (1931) 485.

Scandent shrub or extensive liana. Bark usually light chocolate-brown, smooth. Branchlets subterete, slightly puberulent soon becoming glabrous, nodes remote. Leaves opposite, chartaceous, green, petiole flexuous, 1–2.5 cm long, sparsely puberulent, apically biglandulose, broadly ovate-lanceolate, basally rounded or cuneate-decurrent, apically acuminate-cuspidate (up to 2 cm), marginally entire, 8–18 cm long, 3.5–8 cm wide, upper surface glabrous, nether surface glabrous or with sparse adpressed hairs. Inflorescences axillary or terminal cymose panicles, shorter than or equal to leaves, foliaceous bracts ovate, acute, basally rounded, pubescent, 1.5–3 cm long, 1–1.5 cm wide; basally fleshy-glandular, axis and branchlets ashy-villous-tomentellous. Flowers 12–14 mm in diameter, pedicellate, pedicels 3–8 mm long, cinereous-villous, sepals lanceolate-ovate, rather obtuse, densely cinereous-villous, 2.5–3 mm long, 1.5 mm wide, with or without 8 basal glands. Petals pink, cochleate-suborbiculate or ovate, fimbriate, 5–7 mm long, 4–5 mm wide, claw 1.5 mm long, (fifth petal shorter, flat, subelliptic, margin glandular-fimbriate, 2–3 capitate-stipitate glands, claw more robust 2–2.5 mm long). Stamens 10, filaments rather thick, complanate, anthers unequal, elliptic, 0.6–1.2 mm long, more or less pilose, connective glandular-fleshy. Ovary hardly villose, style glabrous, basally thickened, 3 mm long. Samara nut 5 mm long, adpressed-villous, venulose-costate, basally acutely verrucose, dorsal wing puberulent or glabrous, semi-ovate or semi-ovate-oblong, upper margin thickened, basally slightly and shortly gibbous, lower erose-crenate, basally not constricted, 2.5–3.5 cm long, 1.2– 1.4 cm wide.

Known from the western Amazon basin of Brazil, Bolivia, Colombia, Ecuador and Peru, wild and cultivated in Indian garden plots. It is probably this species which is cultivated by Indians on the Pacific cost of Colombia and Ecuador.

Other species employed are as follows:

Banistriopsis inebrians Morton, Journ. Wash. Acad. Sci. 21 (1931) 485.

Differing from *B. caapi* mainly in leaf texture, placement of glands and shape of leaves and samaras. Known from the western-most Amazon of Colombia, Ecuador and probably Peru.

Banisteriopsis rusbyana (*Ndz.*) *Morton,* Journ. Wash. Acad. Sci. 21 (1931) 487.

An incompletely understood species, reported from the Amazon of Bolivia, Colombia, Ecuador and possibly Peru. The extraordinary chemical difference between this species and *B. caapi* and *B. inebrians* suggests that thorough chemotaxonomic studies might cast doubt even on its generic relationships.

Tetrapteris Cavanilles

The genus *Tetrapteris* comprises some 90 species of vines, rarely treelets, distributed throughout the humid tropics of the New World.

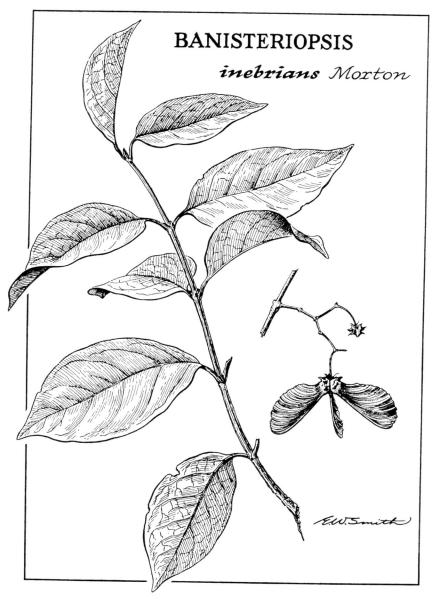

BANISTERIOPSIS

inebrians Morton

Figure 30

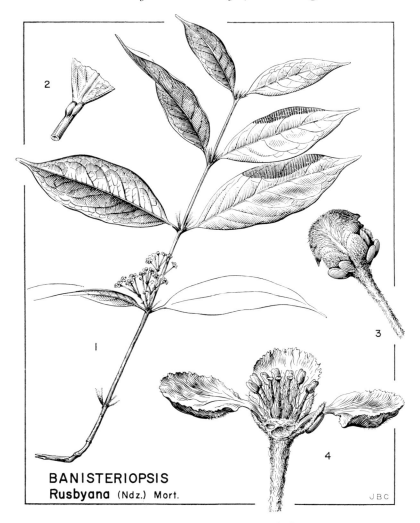

BANISTERIOPSIS
Rusbyana (Ndz.) Mort.

JBC

Figure 31. Drawn by J. B. Clark.

Several writers, notably Spruce and Koch-Grünberg, mentioned more than one "kind" of caapi in the basin of the Vaupés River of Brazil and Colombia.[202,382]

In 1948, Schultes witnessed the preparation and experimented with a narcotic drink amongst nomadic Makú Indians on the Rio Tikié, an affluent of the Uaupés in northwesternmost Brazil.

Figure 32. *Tetrapteris methystica* in forest, Rio Tikié, Amazonian Brazil. Photograph by R. E. Schultes.

Specimens taken from a flowering vine, from the bark of which a cold-water infusion was made without the admixture of any other plants, represented an undescribed species of the malpighiaceous genus *Tetrapteris*, closely allied to *Banisteriopsis*. The species was described and named *T. methystica*.[326]

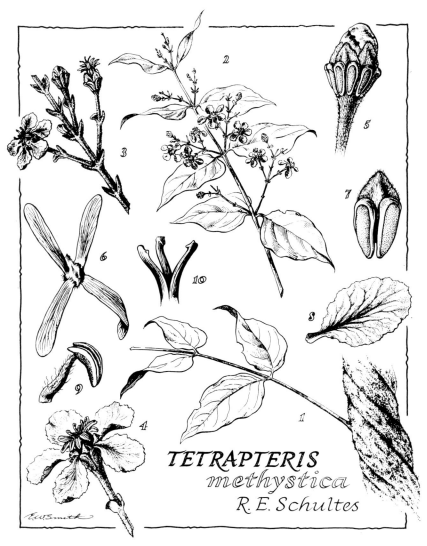

Figure 33

The drink prepared from this jungle liana was of a yellowish hue, quite unlike the coffee-brown colour characteristic of all preparations made from the bark of *B. caapi* in the same region.[326] Unfortunately, chemical studies have not been carried out on

T. methystica. The closeness of the genus to *Banisteriopsis* and the similarity of effects to those induced by *B. caapi* suggest the probability that β-carboline bases may be responsible for its activity.

It is worth considering that *T. methystica* may represent the second "kind" of caapi mentioned by Spruce and Koch-Grün-berg, and it might be that the epithet *caapi-pinima* ("painted caapi") reported by both of these writers alludes not to the coloured leaves but to the unusual yellowish hue of the drink prepared from it.

Tetrapteris methystica *R. E. Schultes,* Bot. Mus. Leafl.
Harvard Univ. 16 (1954) 202.

Scandent bush; trunk with black bark. Branches ashy-yellowish, with internodes 4–10 cm long. Branchlets terete, lightly canaliculate, grey-sericeous when young, 0.8–3.3 mm in diameter. Leaves chartaceous, ovate, rather long acuminate, basally mostly rounded, marginally entire but slightly revolute, 6–8.5 cm long, 2.5–5 cm wide, strongly discolourous, upper surface bright green, minutely and remotely sericeous, nether surface ashy green, rather densely sericeous and provided with wax. Stipules small, soon caducous. Inflorescences pseudocorymbose, few- (apparently 4–5) flowered, much shorter than leaves, 2.5–3 cm long, pedicels rather densely sericeous. Bracts subulate, 1.5 mm long; bracteoles ovate-triangular or suborbicular, 1.5 mm long. Sepals thick, pilose without, ovate-lanceolate, subacute, 3 mm long, with 8 black oval-shaped glands 0.5 mm long. Petals spreading, membranaceous, mostly yellow but red or brownish in centre, elongate-orbicular or oval, apically rounded, basally obtuse, marginally subcrenulate, 4 mm long, mostly 2.5 mm wide, claw fleshy, 0.5 mm long. Stamens not included, equal; anthers allantoid, 1.3 mm long, 0.4 mm in diameter, arcuate, filaments flattened, 1.3 mm long. Styles equal, recurved. Ovary densely albopilose. Samara nut sericeous, glabrescent, complanate-ovoid, 5 mm 6 mm 2 mm, ventral areole ovate, about 3 mm tall; wings chartaceous, brownish, lower ones obcuneiform, apically truncate-rotundate, 12 mm 2.5 mm, upper ones similar but often subovate or semiorbicular and slightly larger; dorsal wing 2.5–5 mm long.

Known only from one locality in the Rio Uaupés basin near the border between Brazil and Colombia.

CACTACEAE

The Cactaceae represents a family of rather enigmatic position and relationships in the Archichlamydeae. It has historically been placed in an order by itself—Opuntiales (Cactales), an order thought to be derived from the Parietales. Recent studies, however, suggest that the family should be assigned near or within the order Centrospermae. One phylogenist hypothesized that the Cactaceae may be derived from the Phytolaccaceae and that the family developed in a kind of parallel evolution with the Aizoaceae. Several taxonomists surprisingly favor even close relationship between the Cactaceae and the metachlamydeous family Cucurbitaceae. There is as yet no agreement. As phytochemical studies of the Cactus Family progress, perhaps chemotaxonomic relationships may aid in a clarification of this mystery.

The Cactaceae comprises about 50 genera (although some systematists recognize from 150 to 220) and perhaps as many as 2000 species. It is native to the drier regions of tropical America (except for one genus, *Rhipsalis,* which occurs, possibly as an early introduction, in Africa and other southern parts of the Eastern Hemisphere), but a few species reach north into Canada and south to Patagonia, as well as up to 11,000 feet in the Andes. Adaptation to extreme xerophytism characterizes the family. Notwithstanding unusual diversity in form—the basis for the excessive variation in taxonomic treatments—floral structure exhibits unexpected similarity throughout the family. Three tribes (sometimes considered subfamilies) are usually recognized: Pereskieae, with broad flat leaves; Opuntieae, with leaves usually more or less terete, small, deciduous; Cacteae, with leaves rudimentary, reduced to minute scales.

Economically, the Cactus Family is important primarily as a source of curious ornamentals. The fruits of a few species enter tropical markets as food. Although aborigines have many uses for cactaceous plants, especially in folk-medicine, there are really relatively few economic species. The family does appear, however, to be a promising repository of highly interesting secondary organic constituents and certainly merits more intensive chemical investigation.

Lophophora Coulter

A genus possibly of two species, allied to *Ariocarpus* and *Mammillaria,* native to southeastern United States and central and northern Mexico.

One of the most important native religions practiced by the pre-Columbian Mexican Indians centered around the worship of the deities through a small, grey-green, napiform, spineless cactus: peyote or *Lophophora williamsii*.[206] It grows in the deserts of central and northern Mexico and adjacent United States and is concentrated especially in the area near the Rio Grande valley.[50]

This cactus, called *peyotl* in Nahuatl as spoken in the Aztec empire, might logically be called the "prototype" of New World hallucinogens, since it was one of the earliest discovered and was

Figure 34. Earliest known botanical illustration of *Lophophora williamsii*. *Botanical Magazine* (1847) t. 4296.

undoubtedly the first and most spectacular vision-inducing plant encountered by the Spanish conquerors of Mexico. The peyote religion was firmly established at the time of the Conquest. It has withstood four centuries of civil and ecclesiastical opposition and is still practiced by a number of Mexican tribes, notably the Tarahumare, Huichol and Cora.[299]

Beginning about 1880, peyote began to spread amongst Indians of the United States, especially certain tribes of Plains Indians. It had been encountered on raids or visits to northern Mexico. The American Indian quickly adopted this purely Indian "sacrament" and built a new ceremony around its use, incorporating indigenous and pagan with Christian elements. The religion spread fast, because of peyote's psychoactive properties and its reputation as a supernatural "medicine." Legally organized now as the Native American Church, the Peyote Cult is known to many Indian groups in the United States and Canada and claims some 250,000 adherents.[209,318,320]

We cannot say how far into the past the use of peyote extends, but it must be very old. The earliest European reports of peyote suggested that the Chichimecas and Toltecs were acquainted with it as early as 300 B.C., although the accuracy of this dating depends on the exactness of interpretation of native calendars by these European writers.[299]

Sahagún, who wrote in the middle of the sixteenth century, stated that ". . . *peiotl* . . . is white; it is found in the north country; those who eat of it see visions either frightful or laughable; this intoxication lasts two or three days and then ceases; it . . . sustains them [the Chichimecas] and gives them courage to fight and not feel fear nor hunger nor thirst; and they say that it protects them from all danger." The Chichimecas first discovered and used peyote, he reported. Another ecclesiastical chronicler wrote in 1591 that they ". . . eat peyote, lose their senses, see visions of terrifying sights like the devil and were able to prophesy the future," and he denounced the plant as constituting "satanic trickery."[308]

Hernández, physician to the King of Spain, wrote in his great treatise on medicinal plants of Mexico that "both men and women are said to be harmed by it. . . . Ground up and applied to painful joints, it is said to give relief. Wonderful properties are attributed to this root. . . . It causes those devouring it to be able to foresee and predict things . . . or to discern who has stolen from them some utensil." [150]

Spanish efforts to stamp out the peyote religion drove it into the hills. In the seventeenth century, a detailed description of the rite amongst the Cora was published. In 1760, a Catholic religious manual used in Mexico equated the eating of peyote with cannibalism. [206]

Whereas in northern Mexico, peyote worship comprised usually a long ceremony in which dancing was a major part, the Indians of the United States have basically a standardized ceremony (with slight variations from tribe to tribe) consisting of an all-night ritual, often in a teepee, with singing, chanting, meditation, prayer and usually a short "sermon" by the road-man or leader, ending in the morning with a communal meal. [206,366]

Peyote is almost invariably eaten in the form of the so-called *mescal buttons*, the dried, brown, disk-shaped tops or crowns of the cactus. They most commonly are simply taken into the mouth, softened with saliva and swallowed without mastication, but occasionally Indians may soak the buttons in water and drink the intoxicating fluid. The crown or the chlorophyllous part of the plant (the only part above ground) is cut from the root and desiccated. In this form, they are well nigh indestructible and may be transported long distances without detriment. The Huichol and Tarahumare in Mexico make annual sacred pilgrimages to the peyote fields, often many miles from their homes, to gather the plant. Most Mexican Indians today get their supply from the Huichol. The indestructibility and durability of the dried buttons have made it possible for the peyote cult to spread far from the native range of the cactus, as far north as Canada. Indian peyotists in the United States and Canada now normally get their supplies by post from Texas. [206,303]

Peyote intoxication, caused by one of the most highly complex and variable of all hallucinogenic plants, is characterized especially by indescribably brilliant coloured visions in kaleidoscopic movement. These visual hallucinations, attributed to mescaline, which is one of the many alkaloids in the plant, are often ac-

Figure 35. *Lophophora williamsii* collected near Laredo, Texas. Photograph by R. E. Schultes.

companied by auditory, taste, olfactory and tactile hallucinations. Sensations of weightlessness, macropsia, depersonalization, doubling of the ego, alteration or loss of time perception and other rather unearthly effects are normally experienced. The very real, and often overlooked, difference between peyote intoxication and mescaline intoxication must constantly be borne in mind. Amongst aboriginal users, it is the dried head of the cactus, with its total alkaloid content, that is ingested; mescaline applied orally or by injection is employed only experimentally and then produces the effects of but one of the alkaloids, without the physiological interaction of the other bases present in the crude plant material. Consequently, descriptions of the visual hallucinations found in reports of psychological experiments should not necessarily be too closely equated with the effects experienced by native peyotists in their ceremonies.[201,206,299,366]

Doses amongst Indian users vary greatly, from four to more than 30 buttons. The ensuing intoxication tends to have two phases: a period of contentment and hypersensitivity and one of nervous calm and muscular sluggishness, often accompanied by hypercerebrality and the typical coloured visions. Before the appearance of the visual hallucinations, the subject sees flashes of colour across the field of vision; the depth, richness and saturation of the colours defy description. There seems to be a kind of sequence followed by the visions: geometric figures, to familiar scenes and faces, to unfamiliar scenes and objects, to secondary objects that vary with individual differences or which may be absent.[36,201]

The literature is rich in excellent and detailed descriptions of visual hallucinations from both peyote and mescaline intoxication, and they provide a wealth of data of interest for psychological and psychiatrical research. While the visual hallucination is important in native peyote cults, peyote is revered in great part because of its appeal as a "stimulant" and "medicine." Its supernatural "medicinal" powers, in turn, come from its ability, through the visions, to put man into contact with the spirit world from which aboriginal belief derives illness and even death and to which medicine men turn for diagnosis and treatment.[322,324]

The magico-therapeutic powers of *Lophophora williamsii* are in such wide repute in Mexico that many other plants are confused or related by vernacular name with it: members of the Compositae, Orchidaceae, Crassulaceae, Leguminosae, Solanaceae, not to mention other cactus species. Members of at least seven cactus genera are related to *L. williamsii* in folk medicine and folk lore, including *Ariocarpus*, *Astrophytum*, *Aztekium*, *Dolichothele*, *Obregonia*, *Pelecyphora* and *Solisia*—all popularly classed as "peyotes," either because they have similar toxic effects and may be employed with *Lophophora* or as a substitute for it or because of some resemblance to *Lophophora*.[319,340]

In 1927, the French pharmacologist Alexandre Rouhier published a remarkable, interdisciplinary book on peyote entitled: *La plante qui fait les yeux émerveillés—le peyotl*.[299] This book represents one of the few and certainly one of the earliest attempts to consider an hallucinogen from all aspects, from the historical, anthropological and social through the botanical, pharmacological and chemical.

It is well known that Indians of northern Mexico have valued

as intoxicants cactus plants other than *L. williamsii*. The explorer
Lumholtz wrote, over 70 years ago, that the Tarahumare ascribe

> . . . high mental qualities . . . to all species of *Mammillaria* and
> *Echinocactus,* small cacti, for which a regular cult is instituted. The
> Tarahumare designate several as *hikuli,* though the name belongs
> properly only to the kind more commonly used by them. These plants
> live for months after they have been rooted up, and the eating of
> them causes a state of ecstasy. They are, therefore, considered
> demigods, who have to be treated with great reverence. . . . The
> principal kinds . . . are known to science as *Lophophora Williamsii*
> and *Lophophora Williamsii* var. *Lewinii.* . . . The Tarahumare
> speak of them as the superior hikuli (*hikuli wanamé*) or simply
> *hikuli,* they being the hikuli *par excellence.* . . . Besides hikuli
> wanamé ordinarily used, the Tarahumare know and worship the fol-
> lowing varieties: 1) Mulato (*Mammil'aria micromeris*) [now known
> as *Epithelantha micromeris*]. This is believed to make the eyes large

Figure 36. Alexandre Rouhier.

and clear to see sorcerers, to prolong life and to give speed to the runners; 2) Rosapara. This is only a more advanced vegetative stage of the preceding species—though it looks quite different, being white and spiny. . . . 3) Sunami (*Mammillaria fissurata*) [now called *Ariocarpus fissuratus*]. It is rare, but it is believed to be even more powerful than wanamé and is used in the same way as the latter; the drink produced from it is also strongly intoxicating. Robbers are powerless to steal anything where Sunami calls soldiers to its aid. 4) *Hikuli walula saeliami*. This is the greatest of all, and the name means "hikuli great authority." It is extremely rare among the Tarahumares, and I have not seen any specimen of it, but it was described to me as growing in clusters of from 8 to 12 inches in diameter, resembling wanamé with many young ones around it. All the other hikuli are his servants. . . . All these various species are considered good, as coming from Tata Dios and well disposed toward people. But there are some kinds of hikuli believed to come from the Devil. One of these, with long white spines, is called *ocoyome*. It is very rarely used, and good for evil purposes." [226]

Figure 37. Arthur Heffter.

Even the modern Tarahumare use other narcotic species of cactus in their festivals; besides *Lophophora williamsii,* they utilize *Ariocarpus fissuratus, A. retusus* and *Epithelantha micromeris,* in both of which alkaloids have been found. Another cactus, *Pachycereus pectin-aboriginum,* is still found in Tarahumare country and is said to be employed as a narcotic by these Indians.[109,261,262]

The earliest serious chemical investigations of *L. williamsii* were carried out at the end of the last century by Heffter, who was successful in isolating a number of alkaloids in pure form from mescal buttons. Studying the pharmacological properties of these substances in tests carried out on animals and in heroic experiments on himself, he found that the main alkaloid, which he named mescaline, has the visual hallucinogenic properties which are characteristic of peyote.[135-137] Späth showed that the chemical structure of mescaline is 3,4,5-trimethoxyphenylethylamine, and he was able to produce the alkaloid by synthesis.[374]

Many variations of this synthesis have since been described.[22,33,82,83,90,129,130,198,367,368,404]

Derivatives of mescaline, such as N-methylmescaline and

N-acetylmescaline, have been isolated from *L. williamsii*.[376,377] Other simple phenylethylamine derivatives were later found in small amounts in peyote: tyramine, N-methyltyramine, hordenine (=anhalin), candicine and 3,4-dimethoxyphenylethylamine.[227,233] Using a combined gas chromatography-mass spectrometry technique, further minor components of peyote could be isolated and identified, such as peyonine, which differs from mescaline by the substitution of the primary amino group by a pyrrolidine carboxylic acid group.[193]

$$CH_3O-\text{⟨ ⟩}-CH_2CH_2N$$
$$CH_3O-$$
$$OCH_3 \qquad COOH$$

peyonine

Furthermore, various N-acylated derivatives of mescaline were found by this technique: N-formylmescaline, N-formyl- and N-acetyl 3,4-dimethoxy-5-hydroxy-phenylethylamine, N-(3,4,5-trimethoxyphenylethyl)-succinimide, N-(3,4,5-trimethoxyphenylethyl)-malimide, N-(3,4,5-trimethoxyphenylethyl)-maleinimide; mescalotam and peyoglutam.[190]

$$CH_3O- \qquad\qquad CH_3O-$$
$$CH_3O- \qquad N \qquad CH_3O- \qquad N$$
$$OCH_3 \qquad\qquad OH$$

mescalotam   peyoglutam

A large group of other minor alkaloids of *L. williamsii* belongs to the structural type of the tetrahydroisoquinoline alkaloids isolated already by Heffter and other investigators.[42,192] These alkaloids are listed in Table IV.

Recent investigations of Kapadia and others have led to the discovery of a large number of derivatives of the alkaloids listed in

TABLE IV

MINOR ALKALOIDS FROM *L. WILLIAMSII* OF THE
TETRAHYDROISOQUINOLINE TYPE

R_1	R_2	R_3	R_4	R_5	*Alkaloid*
H	H	OCH_3	OCH_3	OH	Anhalamine
H	H	OCH_3	OCH_3	OCH_3	Anhalinine
CH_3	H	OCH_3	$O—CH_2$	$—O$	Anhalonine
CH_3	H	OCH_3	OCH_3	OH	Anhalonidine
CH_3	H	OCH_3	OCH_3	OCH_3	O-Methyl-anhalonidine
H	CH_3	OCH_3	OCH_3	OH	Anhalidine
CH_3	CH_3	OCH_3	OCH_3	OH	Pellotine
CH_3	CH_3	OCH_3	$O—CH_2$	$—O$	Lophophorine

this table. N-ethylanhalonine was isolated in minute quantities
and named peyophorine.[191] Anhalidine, lophophorine and pellotine were found in the form of quarternary ammonium alkaloids
and named anhalotine, lophotine and peyotine respectively.[194]
Furthermore N-acyl derivatives of the tetrahydroisoquinoline
type peyote alkaloids were shown to occur in small amounts:
N-formyl anhalinine, N-formyl-O-methyl anhalonidine, N-formyl-
and N-acetyl-anhalamine, N-formylanhalonidine, N-formyl- and
N-acetylanhalonine.[190]

No reports are available concerning hallucinogenic activity for
this large number of minor peyote constituents. It is mescaline
which seems to be mainly responsible for the visual hallucinogenic property of *L. williamsii.*

Mescaline for the first time offered the possibility of producing
and studying the phenomenon of visual hallucinations with a
pure chemical compound. It was, in effect, the first chemically
pure hallucinogen. An average oral dose in man is 0.2 gm, whereas
0.6 gm constitutes a relatively high dose.

Usually the mescaline intoxication starts with unpleasant
autonomic symptoms: nausea, tremor, perspiration. These sub-
side after one to two hours and are replaced by a dreamlike
hallucinatory state, lasting from 5 to 12 hours. The symptoms
typical of a hangover precede the characteristic mental effects.

A review on the pharmacology of mescaline has been given by Fischer.[102] In man, it causes a syndrome of central sympathetic stimulation similar to that of LSD and psilocybin, characterized by increased pupillary size, increase in pulse rate and blood pressure, and elevation of body temperature. Mescaline also decreases the threshold for elicitation of the knee jerk.[445]

Many descriptions of the mescaline-induced mental state are available. From the vast amount of literature concerning clinical reports on peyote- or mescaline-intoxication and the possibilities of its being used in psychiatry, the classical publications of Prentiss and Morgan,[280] Mitchell,[239] and Ellis [89] as well as the monographs of Lewin,[218] Beringer,[36] and Klüver[201] should be mentioned.

Lophophora williamsii (*Lem.*) *Coulter,* Contrib. U. S. Nat. Herb. 3 (1894) 131.

Plant simple, rarely caespitose, normally unicephalous but becoming polycephalous with age or injury, spineless, very succulent, dull bluish or greyish green; roots napiform, usually 8–11 cm long. Crowns globular, top-shaped or somewhat flattened, 2–8 (usually 5–6) cm in diameter, with 7–13 (rarely fewer or more) broad, rounded, straight or spiralled, sometimes irregular and indistinct, with transverse furrows forming more or less regular, polyhedral tubercles; areolas round, flat, bearing flowers only when young, with tuft of long erect, matted woolly hairs. Flowers solitary, borne at umbilicate centre of crown, each surrounded by mass of long hairs, usually pale pink (rarely whitish), rotate-campanulate, 1.5–2.5 cm across when open; outer perianth segments and scales ventrally greenish, callous-tipped; filaments shorter than perianth segments, stigma-lobes 5–7, linear, pink, ovary naked. Fruit club-shaped, red to pinkish, 2 cm long or shorter. Seeds 1 cm in diameter, with broad basal hilum, tuberculate-roughened.

Known from central Mexico north to southern Texas and New Mexico, growing isolated or in groups usually in calcareous deserts, on rocky slopes or in dried river beds.

Many binomials and trinomials have been proposed for *L. williamsii,* due partly to disagreement concerning generic

Figure 38. *Lophophora williamsii*. Jim Hogg County, Texas. Photograph by D. S. Correll.

boundaries and partly to the great variation in number and form of the ribs of the crown. Some systematists have tended to accept as specific or varietal concepts variations which are simple age-forms.[47] There is every intergradation between these apparently very distinct age-forms, and in some instances it is possible to find several of these so-called "species" or "varieties" growing from one root!

The technical names most commonly employed in chemical and pharmacological literature for *L. williamsii* are *Echinocactus williamsii, Anhalonium williamsii, E. lewinii, A. lewinii, A. williamsii* var. *lewinii.*

Figure 39. *Lophophora williamsii* in Huichol Indian territory, Chihuahua, Mexico. Photograph by P. T. Furst.

Inasmuch as the nomenclature and taxonomic treatment of the peyote cactus are so confused, and since a knowledge of this confusion is directly related to a clear understanding of the chemistry of the plant, it will be well to append here a brief and simplified discussion of these botanical points.

The true peyote cactus was first described botanically by Lemaire in 1845 and was assigned to the genus *Echinocactus* as *E. williamsii*. In 1872, Voss considered that it represented a species of *Ariocarpus* and made the appropriate nomenclatural transfer: *A. williamsii*. Shortly thereafter, in 1885, Lemaire himself transferred this species-concept to the genus *Anhalonium*, and the binomial *Anhalonium williamsii* persisted for many years in the anthropological and chemical literature. In 1891, Coulter included this plant in the genus *Mammillaria,* but three years later he described a monotypic genus, *Lophophora,* to accommodate this anomalous species and transferred it as *L. williamsii.*

One of the synonyms most widely encountered in the chemical literature is *Anhalonium lewinii.* It is generally believed that the earliest chemical investigations of the peyote cactus were undertaken by Parke, Davis & Company on material sent in from Laredo, Texas, by a Mrs. Anna B. Nickels. It is known, however, that the German toxicologist, Lewin, acquired material of peyote during his travels across the United States in 1886 and that he submitted the material for botanical study to Hennings, who described it as a new species of *Anhalonium,* naming it *A. lewinii.*[179] In 1888, Lewin ascertained that this material had alkaloids—apparently the earliest record of alkaloids in the Cactaceae.[216]

Anhalonium lewinii was thought to differ morphologically from what had up to then been called *A. williamsii,* the initiation of a confusion that has confounded botanical and phytochemical research for many years. Hennings had received dried mescal buttons which he soaked in water to describe, and he crudely figured the plant from this material.[149] It is now apparent that Hennings was describing merely an age-phase of the cactus. He based his distinction of *A. lewinii* on such characters as number of ribs and tubercles and differences in whiteness, length and silkiness of the tufts of hair. There seemed at the time, furthermore, to be differences in chemical content. Even Hennings, however, confessed that he could not recognize from morphological characters whether he had *A. williamsii* or *A. lewinii,* but he insisted that he could distinguish the two concepts chemically. In the years that followed, the epithet *lewinii* was bandied about: used as a species of *Lophophora* (*L. lewinii*), or as a variety of *L. williamsii* (*L. williamsii* var. *lewinii*), and even as a species of *Mammillaria* and *Echinocactus.*

Recent taxonomic studies suggest that there may be a second species of *Lophophora, L. diffusa,* restricted to the State of Querétaro in central Mexico.[17a.,51a.] It has been postulated that *L. diffusa* may represent the ancestral type in the genus.[42a.] Chemical differences—the presence or absence of mescaline, lophophorine and pellotine—have been associated with morphological characters to suggest the existence of two major populations.[399a.]

Trichocereus Riccobono

The genus *Trichocereus,* a segregate from *Cereus,* comprises some 40 species of subtropical and temperate areas of South America, especially in the Andean regions.

Curiously, the natives of South America have discovered the hallucinogenic properties of a large columnar cactus, not closely related to peyote, in which mescaline has been found as the active principle. The plant, the basis of a psychotomimetic preparation employed in eastern Ecuador and Peru and known locally as *San Pedro,* was misidentified in earlier research reports as *Opuntia cylindrica* [125,126] but is now known to represent. *T. pachanoi.*[106]

The *Trichocereus* may be employed alone or may form the basis of an hallucinogenic beverage called *cimora* and containing,

Figure 40. *Neoraimondia macrostibas.* From N. L. Britton and J. N. Rose, *The Cactaceae 2* (1920) Figure 257. Carnegie Institution of Washington, Washington, D. C.

amongst several other plant ingredients, the cactus *Neoraimondia macrostibas*, the amaranthaceous *Iresine*, a species of *Datura*, the euphorbiaceous *Pedilanthus tithymaloides* and the campanulaceous *Isotoma longiflora*. The drink is commonly taken by witch doctors for diagnosis of disease, divination and to make oneself owner of another's identity.[70]

Mescaline has been isolated from *T. pachanoi* in good yield (2% of dried material or 0.12% of fresh plant). In addition, 3,4-dimethoxyphenylethylamine and 3-methoxy-tyramine were found in essential quantities as well as trace amounts of hordenine, tyramine, 3,5-dimethoxy-4-hydroxy-phenylethylamine, 3,4-dimethoxy-5-hydroxy-phenylethylamine and anhalonidine which were identified by mass spectroscopy.[8,274,347,349] N,N-dimethyl mescaline was found in *T. terschekii*.[292]

Trichocereus pachanoi Britton & Rose, The Cactaceae
2 (1920) 134, t. 196.

Plant 9–20 feet in height. Branches strict, glaucous when young, dark green in age. Ribs 6 to 8, basally broad, obtuse, with deep horizontal depression above areole. Spines few, 3 to 7, often not present, unequal, up to 1–2 cm long, brownish. Buds pointed.

Figure 41. *Trichocereus pachanoi* in a native market place. Andean Peru. Photograph by C. Friedberg.

Flowers large, 19–23 cm long, borne near apex of branches, night-blooming, very fragrant, outer perianth segments brownish red, inner segments white; filaments of stamens long, greenish; style greenish below, white above; stigma lobes linear, yellowish; ovary black-pilose. Axis of scales on flower-tube and fruit with long black hairs.

This species of *Trichocereus* occurs in the Andean parts of Ecuador and Peru and probably in Bolivia, between 6,000 and 9,000 feet. It is apparently widely cultivated throughout the central Andes.

LYTHRACEAE

The Lythraceae, belonging to the order Myrtiflorae, comprises about 25 genera and perhaps some 580 species of herbs, shrubs and trees of the temperate and tropical parts of both hemispheres. The family is economically noteworthy as the source of ornamentals and of the dye henna.

Heimia Link & Otto

A genus of three hardly distinguishable species, *Heimia* ranges from southern United States to Argentina.

Sinicuichi is an interesting and still poorly understood narcotic drink of the Mexican highlands. The leaves of *H. salicifolia*, slightly wilted, are crushed in water, and the juice is set out in the sun to ferment. The resulting beverage is a mild intoxicant causing a slight giddiness, a darkening of surroundings, shrinkage in size of the world around, a pleasant drowsiness or euphoria, deafness or auditory hallucinations with distorted sounds coming apparently from great distances. There are usually no uncomfortable after effects. Excessive use of the drink is said to be harmful.[290]

More ethnobotanical field work is needed before fact can be separated from fancy in the case of sinicuichi, but there is no doubt that it is employed—albeit not ritually nor ceremonially—as an hallucinogen. Natives of Mexico believe that sinicuichi has sacred or supernatural virtues. They assert that it helps them remember events that took place many years earlier as though they had happened yesterday and that they are able, through the plant, to recall even prenatal events or conditions.[409]

Heimia salicifolia and the very closely allied *H. myrtifolia* (which may represent only a geographical variant) range in the highlands from Mexico south to Uruguay, Paraguay and northern Argentina. Throughout this range, these plants enjoy unusual uses in folk medicine, but apparently only in Mexico are the hallucinogenic properties important in native practices. Some of the local vernacular names suggest biodynamic properties: in Brazil, *abre-o-sol* ("sun opener") and *herva da vida* ("herb of life"). In Mexico, the term *sinicuichi* (or derivatives of it such as *sinicuil* or *sinicuilche*) refers to other plants as well, but all of the plants so denominated are, in one way or another, intoxicating (e.g. species of *Erythrina, Piscidia and Rhynchosia*).[347,349]

A review of ethnological, chemical and pharmacological aspects of *H. salicifolia* was published in 1966 by Tyler.[409]

The first systematic investigation on the alkaloids of *H. salicifolia* was done in 1964 by Blomster and others,[40] who isolated and described four alkaloids: lythrine, cryogenine, heimine and sinine. The isolation of two additional alkaloids, vertine and lythridine was reported by Douglas and others.[84] Finally, Appel and others [18] isolated two additional alkaloids, which they named nesodine and lyofoline. Further investigations showed that cryogenine may be identical with vertine. The complete structure of the lythraceous alkaloids which belong to the quinolizidine type was elucidated by Zacharias and others [451] and by Ferris and others,[100] based on x-ray analysis.

Cryogenine (Vertine)

Of these alkaloids, cryogenine (vertine) seems to possess the most significant pharmacological activity, mimicking qualitatively and quantitatively the action of the total alkaloids of the plant.[297,298] In the hippocratic screen, cryogenine was found to possess antispasmodic, anticholinergic, skeletal muscle relaxing and tranquilizing properties. Whereas a detailed picture of the pharmacological effects of cryogenine and of other alkaloids of *H. salicifolia* is available,[195] no reports exist dealing with their effects in man which could be related to the hallucinogenic properties reported after ingestion of preparations of the whole plant.

Heimia salicifolia (*HBK.*) *Link*, Enum. Pl. 2 (1822) 3.

Shrub 2–6 feet in height, glabrous throughout. Leaves mostly opposite, some in threes or uppermost alternate, sessile, linear-lanceolate or lanceolate, acute or acuminate, 2–9.5 cm long. Flowers solitary in axils, yellow, without aroma, pedicellate. Calyx campanulate, about 5–9 mm long, with long hornlike appendages at base of lobes. Petals soon caducous, 5 to 7, obovate, 12–17 mm long. Stamens 10–18.

Common along streams and in moist places in the highlands of Mexico and western Texas, El Salvador, Jamaica and northern South America.

APOCYNACEAE

The Apocynaceae is a natural family in the order Contortae, closely related to the Asclepiadaceae, Loganiaceae and Gentianaceae. Comprising some 180 genera and at least 1700 species, mostly twining shrubs or lianas, rarely trees, the family is pre-eminently tropical but has a few temperate representatives. Almost all—if not all—species contain a sticky, white latex. The family is divided usually into two sections: Plumieroideae, with three tribes; and Echitoideae, with two tribes. A few genera are economically important, mainly as sources of rubbers and drugs.

One of the mysteries in the study of narcotics employed in primitive societies is why the Apocynaceae, probably the family richest in alkaloids, should be so sparingly represented in the list of species valued and utilized for their psychotomimetic proper-

HEIMIA
salicifolia
Link & Otto

Figure 42. Drawn by I. Brady.

ties. There are undoubtedly sundry species in this family possessing organic constituents capable of inducing visual or other
hallucinations, but either they have not been discovered by
aborigines or they are too toxic for human consumption. It is
possible, too, that future ethnobotanical field work will uncover

euphorbiaceous *Alchornea floribunda.* This plant is said to be employed in the same way as iboga in another secret society in Gabon—the Byeri—and may well be an hallucinogen.

The chemistry of *A. floribunda* is not yet elucidated. Earlier findings that the roots and seeds of this plant contained yohimbine[258] have not been confirmed.[283] The alkaloids which were found still remain to be identified.[409]

Tabernanthe iboga *Baillon*, Bull. Soc. Linn. Paris 1 (1889) 783.

Small shrub up to 6 feet in height. Root robust, much branched. Branches slender, terete, lenticillate. Leaves petiolate (petiole 2–3 mm long), elliptic-ovate, or obovate-lanceolate, acuminate, basally acute or long cuneate, mostly 7.5–13 cm long, 2.5–4.5 cm wide; nerves 9–11, oblique, arcuate. Inflorescence loosely umbelliform or subcorymbose, few- to about 12-flowered, nutant, shorter than leaves; peduncle 1–4 cm long; pedicels about 8 mm long. Flowers white, sometimes with pink spots; calyx 5-partite, 1–1.5 mm long; sepals broadly ovate or subtriangular, ciliolate, the inner with 1 or 2 basal glands within; corolla subcylindric, narrowed from middle up, about 5 mm long, lobes rotundate, 2.5 mm long; stamens inserted at middle of tube; anthers 2 mm long; style 2 mm long. Fruit ellipsoid, 18–24 mm long with smooth, crustaceous pericarp, sometimes crowned with persistent base of style. Seeds globose or somewhat ellipsoidal, 6 mm long with corky, lamellate-rugose testa.

Native to Gabon and the southeastern parts of the Congo in Africa.

Alchornea floribunda *Mueller-Argoviensis*, Fl. Ratisb. (1864) 435.

A species that occurs across equatorial Africa.

CONVOLVULACEAE

Ipomoea Linnaeus; *Rivea* Choisy

Ipomoea is a large genus of more than 500 species of the tropical and warm-temperate parts of both hemispheres, almost all of the species being climbing herbs or shrubs. The genus has several species of medicinal value as the sources of purgatives, but the most important economic representative is undoubtedly the sweet potato (*I. batatas*).

Rivea has been separated from *Ipomoea* on several minor characters which often intergrade, and there may be justification for including all species once again in *Ipomoea*. *Rivea* comprises half a dozen species, five native to India and southwestern Asia and one (*R. corymbosa*) tropical and subtropical America.

Numerous early chroniclers of the time of the Spanish conquest of Mexico reported the religious and medicinal use of a small lentil-like, hallucinogenic seed called *ololiuqui*. The Aztecs and other Indians ingested ololiuqui for purposes of divination. The seed came from a vine with cordate leaves known in the Nahuatl language as *coaxihuitl* or "snake-plant." [150,308]

In 1615, Ximénez published some of Hernández's ethnobotanical notes. Without identifying ololiuqui, he stated that "it will not be wrong to refrain from telling where it grows, for it matters little that this plant be here described or that Spaniards be made acquainted with it." [449] A number of early references and several illustrations, however, indicated that it was convolvulaceous. Hernández described and figured it. Sahagún, Hernández's contemporary, enumerated three plants called ololiuqui, one of which was ". . . an herb called *coatlxoxouhqui*, and it bears a seed called ololiuqui." The *Florentino Codex* illustrated clearly a convolvulaceous vine with cordate leaves and a tuberous root.

An early record, written in 1629, reported that ". . . when drunk, the seed deprives of his senses him who has taken it, for it is very powerful." Still another source said that "it deprives those who use it of their reason. The natives . . . communicate with the devil . . . when they become intoxicated with ololiuqui, and they are deceived by the various hallucinations which they attribute to the deity which they say resides in the seeds. . . ." A further report described how ". . . these seeds . . . are held in great veneration. . . . They place offerings to the seeds . . . in secret places so that the offerings cannot be found. . . . They also place these seeds amongst the idols of their ancestors. . . . They do not wish to offend ololiuqui with demonstrations before

Figure 45. *Rivea corymbosa* or ololiuqui of the ancient Mexicans. From F. Hernández, *Rerum medicarum Novae Hispaniae thesaurus, seu plantarum, animalium, mineralium mexicanorum historia.* Rome (1651).

the judges of the use of the seeds and with public demonstrations of the seeds by burning." [325]

Ololiuqui was used also as a magic potion with reputedly analgesic properties. This is an unexplained aspect of the ethnopharmacology of the seeds. One report asserted that the Aztec priests, before making sacrifices on mountain tops, ". . . took a large quantity of poisonous insects . . . burned them . . . and beat the ashes together . . . with the foot of the *ocotl*, tobacco, ololiuqui and some live insects. They . . . rubbed themselves with this diabolic mixture and . . . became fearless to every danger." Another recorded that ". . . this unction was made of diverse little venomous beasts . . . with much tobacco or pectum . . . , an herb that they use much to benumb the flesh . . . then they put to it a certain seed . . . called *ololuchqui,* whereof the Indians make a drink to see visions. . . . The priests being slobbered with this ointment lost all fear . . . they said that they felt thereby a notable ease, which might be, for that the tobacco and *ololuchqui* have this property of themselves, to benumb the flesh." Hernández, whose report seems to be the most reliable of the early writers, mentioned its presumed pain-dulling effects and, after a detailed description of its many medicinal uses, stated that "when the priests wanted to commune with their gods and to receive a message from them, they ate this plant to induce a delirium, during which a thousand visions and satanic hallucinations appeared to them." [325]

For nearly four centuries no convolvulaceous plant was encountered in use in Mexico as a divinatory or ritualistic hallucinogen. Furthermore, no intoxicating constituent had ever been isolated from any convolvulaceous species. This presented an enigma. In 1911, Hartwich suggested that ololiuqui might well be solanaceous,[133] and Safford, in 1915, arguing that the early chroniclers must have been misled by the Indians, definitely identified the narcotic as *Datura meteloides*.[301] His identification was widely accepted in anthropological, botanical and pharmacological literature. There were voices of protest. B. P. Reko, for example, in 1919 accepted Urbina's identification of ololiuqui as *R. corymbosa* (*I. sidaefolia*), made in 1897 and reiterated in 1903 and 1912; [287,411,412] in 1934, Reko published an historical review of ololiuqui.[289] It was not, however, until 1939 that Schultes and Reko collected identifiable botanical material of *R. corymbosa* from a cultivated plant employed in divination by a Zapotec witch doctor in northeastern Oaxaca.[325] The seed is used amongst other Indians in Oaxaca—Chinantecs, Mazatecs, Mixtecs—and, as Wasson states, "to-day in almost all the villages of Oaxaca one finds seeds still serving the natives as an ever present help in time of trouble." [429]

Although toxic principles were unknown from the Convolvulaceae, it was Santesson who, in 1937, first reported psychoactivity from seeds of *R. corymbosa*.[311,312] He was unsuccessful, however, in isolating definite crystalline compounds. Alcoholic extracts produced a kind of narcosis or partial narcosis in frogs and mice, and certain chemical reactions seemed to suggest to him the presence of a gluco-alkaloid.

In 1955, the psychiatrist Osmond conducted a series of experiments on himself, and, after taking 60 to 100 seeds of *R. corymbosa*, he experienced a state of apathy and listlessness, accompanied by increased visual sensitivity. After about four hours, there followed a period in which he had a relaxed feeling of well-being that lasted for a longer time.[255] In contrast to these results, Kinross-Wright, in 1959, reported experiments on eight male volunteers who had taken doses of up to 125 seeds without any ascertainable effect in even a single case.[199]

The enigma of ololiuqui and the chemical nature of its active principles was solved in 1960 by Hofmann. Analysis of seeds of *R. corymbosa* from Oaxaca revealed the surprising discovery that the psychoactive constituents of ololiuqui are ergot alkaloids.[176]

Another step in the study of the narcotic morning glories of Mexico came in 1960 when MacDougall found the seeds of *I. violacea* (*I. tricolor*) taken together with or in place of those of *R. corymbosa*, especially amongst certain groups of Zapotecs in Oaxaca who refer to them as *badoh negro*.[228] These seeds, which are jet black and which have a different shape from those of *R. corymbosa*, had been reported earlier by this vernacular name by Parsons from the Zapotecs of Mitla.[259] Wasson has suggested that this species represents the ancient Aztec narcotic called *tlitliltzin*, a term in Nahuatl derived from the word for "black"

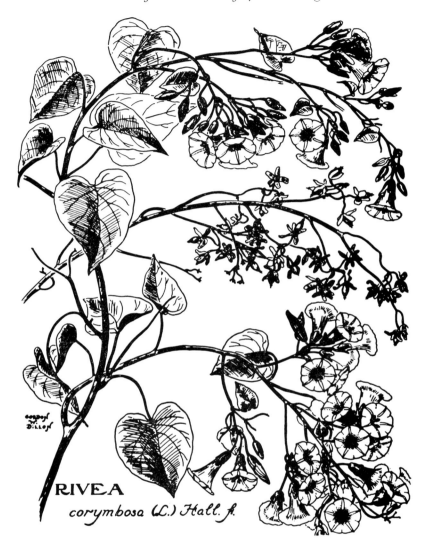

RIVEA
corymbosa (L.) Hall. f.

Figure 46

with a reverential suffix.[429] One of the early chroniclers, for example, had written of "ololiuqui, peyote and tlitliltzin."

Chemical studies of the seeds of *I. violacea* have completely substantiated ethnobotanical data pointing to their utilization as

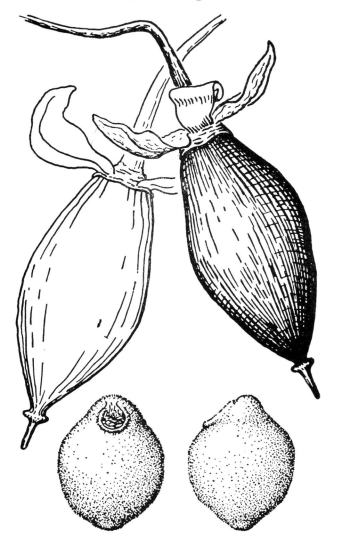

Figure 47. Capsules and seeds of *Rivea corymbosa*, magnified twenty times. Drawn by G. W. Dillon.

hallucinogens. Hofmann reported that, as with R. *corymbosa*, the psychotomimetic principles are ergot alkaloids.[159,161,162,176]

From the phytochemical point of view, this finding was wholly unexpected and of particular chemotaxonomic interest, since

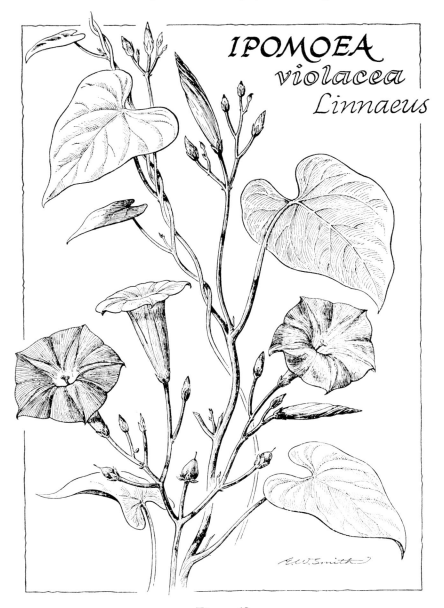

Figure 48

lysergic acid alkaloids, which had hitherto been isolated only from the lower fungi of the genera *Claviceps, Penicillium* or *Rhizopus,* were now, for the first time, found in higher plants—in the phanerogamic family Convolvulaceae. Subsequent chemical investigations in other laboratories confirmed the occurrence of ergot alkaloids in other convolvulaceous species of *Argyreia, Ipomoea* and *Stictocardia.*[37,74,75,124,183,392-394]

D-lysergic acid amide, also called ergine, is the main constituent of ololiuqui: seeds of *R. corymbosa.* Furthermore, the following minor alkaloids were isolated: d-isolysergic acid amide (isoergine), chanoclavine, elymoclavine and lysergol. The seeds of *I. violacea* yielded the same alkaloids, with the difference that ergometrine (syn., ergonovine) was obtained in place of lysergol. Later, it was found that ergine and isoergine were present in the seeds to some extent in the form of d-lysergic acid N-(1-hydroxyethyl) amide and d-isolysergic acid N-(1-hydroxyethyl) amide, respectively, and that, during the isolation procedure, they easily hydrolize to ergine and isoergine, respectively, and acetaldehyde. The presence of further minor alkaloids (ergometrinine and penniclavine) was detected only by chromatographic tests.[161]

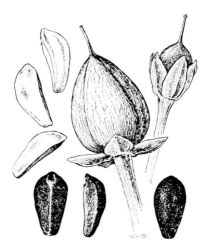

Figure 49. Capsule and seed of *Ipomoea violacea,* enlarged two and one half times. Drawn by E. W. Smith.

TABLE V

STRUCTURAL FORMULAE OF OLOLIUQUI ALKALOIDS

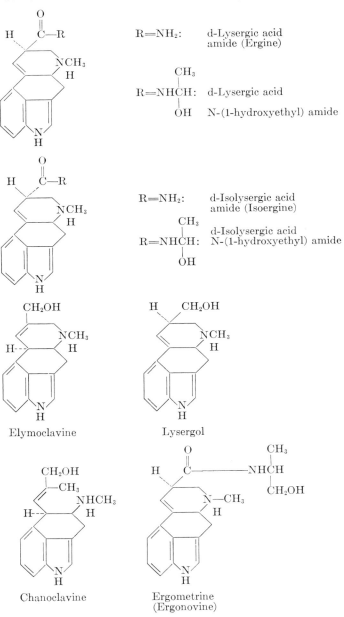

R=NH₂: d-Lysergic acid
               amide (Ergine)

R=NHCH: d-Lysergic acid
               N-(1-hydroxyethyl) amide

R=NH₂: d-Isolysergic acid
               amide (Isoergine)

R=NHCH: d-Isolysergic acid
               N-(1-hydroxyethyl) amide

Elymoclavine   Lysergol

Chanoclavine   Ergometrine
                                (Ergonovine)

The total alkaloid content of seeds of *R. corymbosa* was found to be 0.012 percent, whereas the seeds of *I. violacea* contain 0.06 percent of total alkaloids. This fact explains why Zapotec Indians use smaller quantities of seeds of the latter species than of the former.[161]

d-Lysergic acid amide and d-isolysergic acid amide were first obtained as products of the alkaline hydrolysis of ergot alkaloids,[369,370] then also by partial synthesis from lysergic acid, and more recently, as naturally occurring alkaloids, together with the corresponding N-(1-hydroxyethyl) amides from ergot of *Paspalum*.[19] Chanoclavine has previously been discovered in ergot of the tropical millet, *Pennisetum typhoideum*.[170] Elymoclavine was first isolated from ergot of the wild grass, *Elymus mollis*.[1] Lysergol was produced synthetically by reduction of d-lysergic acid,[386] before it was found to occur in nature as one of the active principles of ololiuqui. Ergometrine (ergonovine) is the alkaloid mainly responsible for the uterotonic haemostatic action of the ergot drug. It can also be obtained synthetically.[385]

The psychotomimetic activity of lysergic acid amide and its marked narcotic component was ascertained from self-experiment by Hofmann.[161] This action of d-lysergic acid amide was later confirmed by comparative systematic investigations by Solms.[373] He described the action as follows: LA-111 induces indifference, a decrease in psychomotor activity, the feeling of sinking into nothingness and a desire to sleep . . . until finally an increased clouding of consciousness does produce sleep.

Only little information is available on the activity of d-isolysergic acid amide (isoergine). After taking 2.0 mg orally, Hofmann experienced tiredness, apathy, a feeling of mental emptiness and the unreality and complete meaninglessness of the outside world.[161]

D-lysergic acid N-(1-hydroxyethyl) amide induces contractions in the isolated uterus of the guinea pig and in the rabbit uterus *in situ*, showing about 30 to 50 percent of the activity of ergometrine. In mice and rabbits, it produces the syndrome of central sympathetic stimulation, such as pilo-erection, mydriasis and hyperthermia, which suggests that it could have an LSD-like activity; however, this hypothesis has not yet been verified by experiments on humans.

Elymoclavine and lysergol elicit an excitation syndrome in various animals, caused by a central stimulation of the sympathetic nerves which seems to indicate psychotomimetic activity.[450] Results of clinical tests are not as yet available.

Psychotomimetic effects are unknown for ergometrine, which is used to a large extent in obstetrics as a uterotonic and haemostatic agent. In small dosages administered for this purpose, the alkaloid apparently has no action on the psychic functions. Its occurrence in the alkaloid mixture of ololiuqui can thus have no significant effects on the mental action.

Furthermore, chanoclavine, which has no outstanding pharmacological activity, appears to play no part in the occurrence of the psychic effects of ololiuqui.

According to the results of experiments performed thus far with pure alkaloids, it appears that d-lysergic acid amide, d-lysergic acid N-(1-hydroxyethyl) amide, elymoclavine and lysergol, and possibly also d-isolysergic acid amide are responsible mainly for the psychotomimetic effects of ololiuqui.

The isolation of lysergic acid amides from *I. violacea* and *R. corymbosa* brings these ancient "magic plants" into direct relationship with LSD-25, the laboratory code name for d-lysergic acid diethylamide. LSD, the most powerful hallucinogen so far known, is a synthetic compound; it differs from the main constituent of ololiuqui—from d-lysergic acid amide—only by the substitution of two hydrogens at the amide group by two ethyl radicals.

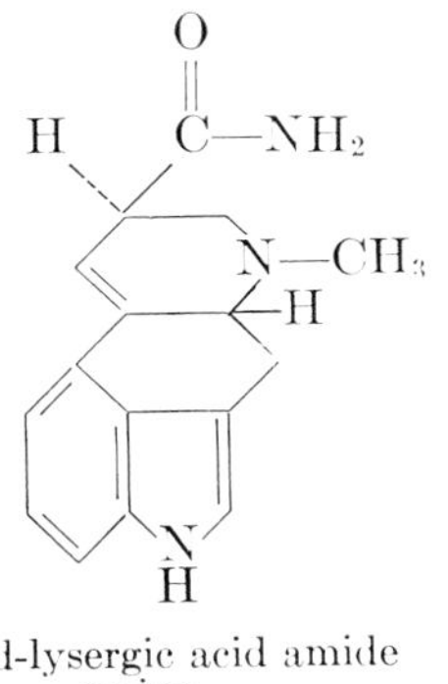

d-lysergic acid amide
ergine
(from ololiuqui)

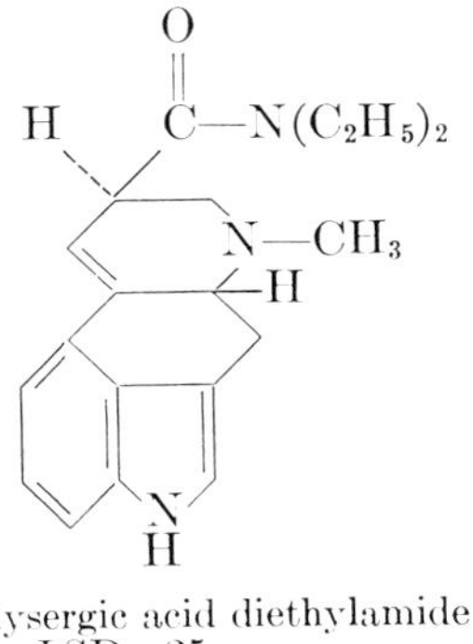

d-lysergic acid diethylamide
LSD—25
(semisynthetic)

their psychoactive properties. It is perhaps of interest to note here that one of the "new" hallucinogens used in the United States in artistic, literary and other "sophisticated" groups devoted to experimentation with drugs is catnip, *Nepeta cataria*, a labiate.[187]

Salvia Linnaeus

There are some 700 species of *Salvia* distributed in the temperate and tropical parts of both hemispheres. The most important economic species is *S. officinalis*, the common garden sage which is employed as a spice.

In southern Mexico, amongst the Mazatec of Oaxaca, crushed leaves of *S. divinorum,* known locally as *hierba de la Virgen* or *hierba de la Pastora,* are valued for use in divinatory rites for their psychotomimetic properties when other more potent hallucinogens, such as mushrooms and morning glories, are unavailable.[428]

This hallucinogen is known to virtually all Mazatec Indians. Many Mazatec families have a private supply growing, almost always away from home sites and trails as if they wished to hide the plant from passersby. The plant is reproduced vegetatively by breaking off a shoot and sticking it into rich black soil at altitudes of about 5500 feet. *S. divinorum* is apparently a cultigen and seems not to occur wild, an indication perhaps of great age in an agricultural or horticultural setting.[428]

Whether or not this hallucinogen is used beyond the Mazatec country is not yet known; the immediately contiguous Cuicatec and Chinantec Indians in Oaxaca may possibly employ it as well. It has been suggested that *S. divinorum* represents the psychotomimetic *pipiltzintzintli* of the ancient Aztec.[428]

Although investigators have experimentally substantiated the psychotomimetic activity of *S. divinorum,* chemical studies have thus far failed to yield a psychoactive principle.[168,347,349]

Salvia divinorum Epling & Jativa-M., Bot. Mus. Leafl., Harvard Univ. 20 (1962) 75.

Perennial herb, 3 feet in height or taller. Leaves 12–15 cm long, ovate, acuminate, basally more or less rounded, crenate-serrate with hairs in sinuses along margins, glabrate but hirtellous along

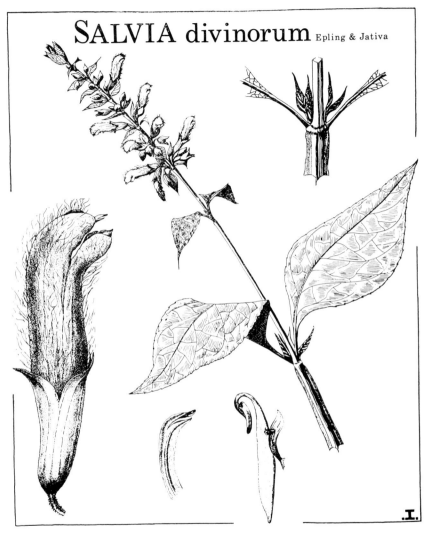

Figure 52. Drawn by I. Brady.

lower veins; attenuated into a petiole 2–3 cm long. Flowers bluish, slightly pubescent, in full panicles on branches 30–40 cm long. Calyx tube bluish, 15 mm long with superior lip 1.5 mm long and 3 impressed veins; sigmoid corolla tube blue, 22 mm long, with superior lip 6 mm tall, inferior lip shorter and incurved; stamens inserted near mouth of tube, included; style hirtellous, with posterior branch rather long, obtuse, flat, anterior branch apparently carinate.

Known only from cultivated material in forest ravines in black soil in northeastern Oaxaca, Mexico, this species is closely related to *S. cyanea* Lindl. of central Mexico.

SOLANACEAE

Belonging to the order Tubiflorae and allied to the Scrophulariaceae and to several other families (suggesting possibly a polyphyletic origin), the Solanaceae has some 90 genera and well over 2400 species of herbs, shrubs, small trees, some of lianous or creeping habit. Of temperate and tropical distribution in both hemispheres, it has its centre of diversification in the Andes of South America. The usually accepted classification breaks the family into five tribes: Nicandreae, Solaneae, Datureae, Cestreae, Salpiglossideae. Economically, the Solanaceae is important as the source of food plants (fruits, starchy tubers, spices), numerous poisons, narcotics and medicines and many ornamentals. One significant characteristic of the family is the prevalence of alkaloids in many species.

Atropa Linnaeus

A genus of four species, distributed in Europe, the Mediterranean area and from central Asia east to the Himalayas.

The *belladonna* plant, *Atropa belladonna*, is well known as a highly toxic species and has been utilized as a poison since early classical times. It is thought that belladonna was often one of the principal ingredients of the hallucinogenic witches' brews in medieval Europe.[133,217,218,418]

The principal active constituent of *A. belladonna* is hyoscyamine, which is accompanied by small amounts of the more specifically psychoactive alkaloid scopolamine (syn. hyoscine). Atropine has been found also in various amounts but may have resulted from racemization of hyoscyamine during extraction. When careful procedures avoiding racemisating conditions are employed, no atropine or only traces can be detected. Other alkaloids which were found also only in trace amounts are tropine, scopine, N-methyl-pyrroline, N-methyl-pyrrolidine, cuscohygrine, belladonnine, and nicotine. Total alkaloid contents were reported in the leaves to be 0.4 percent; in the roots 0.5 percent in the seeds 0.8 percent. There are no pharmacological results supporting hallucinogenic activity of any of the minor components of *A. belladonna* and of the other solanaceous plants. The chemical discussions, consequently, will be limited to the principal constituents with alleged hallucinogenic activity.

Since the first isolation of the principal alkaloids of *A. belladonna* early in the last century, a vast amount of literature concerning chemical structure and synthesis of these compounds has accumulated. Furthermore, the stereochemistry, including the absolute configuration, has been elucidated in recent years, and detailed reviews in this field have been published.[386a,104a,104b] The results of these investigations are summarized in the following formulae:

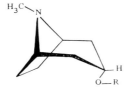   

R=H : Tropine
R=(−)−Tropoyl : Hyoscyamine
R=(±)−Tropoyl : Atropine

R=H : Scopine
R=(−)−Tropoyl:Scopolamine (Hyoscine)

Hyoscyamine is the ester of tropine with (−)-tropic acid, atropine with the racemic (±)-tropic acid. Scopolamine is the (−)-tropic acid ester of scopine. The absolute configuration of (1)-tropic acid is S, as represented by the following formula:

$$CH_2OH$$
$$HOOC—C—H$$

S(−)—Tropic acid

Hyoscyamine and scopolamine possess specific anticholinergic, antispasmodic activity and elicit some central nervous effects as well. These effects usually consist of stimulation in low doses, depression in higher, toxic doses. The anticholinergic activity is due almost wholly to the $S(-)$-form. Hyoscyamine is thus twice as potent as atropine in its antispasmodic activity. In central activity, hyoscyamine is eight to 50 times as potent as the $(+)$-isomer. There are no reports of hallucinogenic effects produced by the pure chemicals atropine or hyoscyamine which could explain the addition of belladonna as an ingredient of magic brews in medieval Europe. Scopolamine, however, was found to produce a state of intoxication followed by a kind of narcosis where, in the transition state between consciousness and sleep, hallucinations sometimes occur.[148a] This accounts for the use of those solanaceous plants containing mainly scopolamine, such as species of *Datura*, for magical purposes.

Atropa belladonna Linnaeus Sp. Pl. (1753) 181.

Branched, perennial herb up to 4½ feet tall. Stems glabrous or viscid-pubescent. Leaves alternate or, especially terminal ones, in unequal pairs, ovate to oblong, 8–20 cm long, entire, acuminate, narrowed into short petiole. Flowers axillary, solitary or rarely paired, 25–30 mm long, on nodding peduncle 1–2 cm long, usually greenish purple. Calyx somewhat accrescent, leafy, deeply 5-fid; lobes triangular-ovate, acuminate. Corolla tubular-campanulate, with 5 broad, imbricated lobes; lobes obtuse. Stamens 5, inserted at base of corolla, included; filaments basally thickened; anther cells distinct. Style exserted. Stigma peltate. Ovary dilated, shortly 2-celled, implanted on disk. Berry globular, many seeded, subtended by spreading calyx, black, 15–20 mm in diameter.

Atropa belladonna is native to Europe.

Brunfelsia Linnaeus

A genus of perhaps 40 species of shrubs of tropical South America and the West Indies.

It seems probable that various species of *Brunfelsia* were once employed as important hallucinogens by sundry Indians of South America. The native names for the plants and the special care and methods that the Indians followed in cultivating them would seem to indicate a former religious or magic role for *Brunfelsia*. Only recently has real evidence of its actual utilization come to the fore. Further ethnobotanical studies should be carried out to ascertain the former and present extent of this hallucinogenic use.[340]

Figure 53. *Brunfelsia grandiflora*. Leticia, Amazonian Colombia. Photograph by T. Plowman.

An obscure and still poorly understood species described as *B. tastevinii* has been reported as the basis of a psychotomimetic drink of the Kachinaua Indians of the Brazilian Amazon, but this report needs confirmation. The Kofán Indians of Amazonian Colombia and Ecuador, as well as the Jivaro of Ecuador, occasionally add the leaves and bark of a cultivated *Brunfelsia* to their yajé or natema drink which is prepared basically from *Banisteriopsis*. This use of *Brunfelsia* and its narcotic properties is well known in the Colombian Putumayo, where the shrub is called *borrachero* ("intoxicant") amongst the non-Indian population.[345,347,349]

The fluid extract of one species—*B. uniflora*—is employed in Brazilian medicine as a diuretic and antirheumatic; it is included in the Brazilian pharmacopoeia. The genus plays an important role in folk medicine with a broad spectrum of therapeutic uses, ranging from the treatment of "yellow fever" to snakebite. Its most widespread folk uses seem to be to relieve "rheumatism" and as a febrifuge, since its ingestion is usually followed by a sensation of chill and coldness, effects that might be explained on the basis of tropane alkaloids.[340]

Only preliminary chemical investigations of species of *Brunfelsia* have been carried out. The older literature mentions the isolation of alkaloidal components given names such as "franciscaine," "manacine" and "brunfelsine," but none of them seems so far to have been satisfactorily characterized. In a more recent report on the investigation of *B. uniflora*, *B. pauciflora* and *B. brasiliensis*, no alkaloids are mentioned, but the isolation of a nitrogen-free compound, scopoletin, (=6-methoxy-7-hydroxy-coumarin), which is not known to be psychotropically active, has been published.[241a]

grandiflora D. Don, Edinb. New Phil. J. (Apr.–Oct. 1829) 86.

Shrub to small tree 1–5 m tall. Bark thin, exfoliating in chartaceous flakes, buff-coloured, somewhat nitid. Leaves alternate, broad-elliptic to oblong-lanceolate, acuminate, entire, basally attenuate, 6–20 mm long, 2–8 cm wide, youngest often pubescent, becoming glabrous, 7–9 lateral veins, arcuate, anastomosing near margin, thin to firm-membranaceous, dark green above, pale green beneath. Petiole 2–6 mm long. Stipules absent. Inflorescence cymose, few- to many-flowered; bracts small, lanceolate, deciduous, 2 mm long. Calyx tubular to somewhat inflated, 5-parted, 5–10 mm long, persistent, glabrous, becoming subcoriaceous in fruit and splitting irregularly; teeth acute, 1–3 mm. Corolla hypocrateriform, slightly zygomorphic, tube 2–3 times longer than calyx, pale violet; the limb undulate, thin-membranaceous, 1.5–4.0 cm across, rounded lobes 1.2 cm long, violet to purple with white ring at throat, fading to white with age. Stamens 4, didynamous, in upper part of tube; outer pair of filaments longer, exceeding pistil; inner pair shorter; the anthers simple, reniform, incumbent, 1 mm long, longitudinally dehiscing. Ovary bilocular, sessile, conical; style incurved, shorter than corolla tube; stigma thick, bilobed, included. Fruit capsular, ovoid to globose, bivalved, bilocular, submucronate, 1.0–1.8 mm in diameter, dark green. Seeds numerous, oblong, 5 mm long, reticulate-pitted, dark reddish brown.

Brunfelsia grandiflora, an extremely variable species, is widely distributed in tropical South America, known from Colombia, Venezuela, Ecuador, Peru, Bolivia and Brazil. It is found in humid forests.

Datura Linnaeus

The genus *Datura* comprises some 15 to 20 species and is divided usually into four sections: (a) *Stramonium,* with three species in the two hemispheres; (b) *Dutra,* containing six species; (c) *Ceratocaulis,* with one Mexican species; (d) *Brugmansia,* South American trees representing possibly six or seven species. *Brugmansia* is sometimes considered a separate genus.

Datura has had a long history in both hemispheres as a genus employed for hallucinogenic purposes.[133,218]

The important narcotic species of the Old World is *D. metel.* Early Sanskrit and Chinese writings report an hallucinogen that has been identified with this species, and it was probably *D. metel* that the Arabian Avicenna mentioned as a drug called *jouz-mathel* in the eleventh century. In 1578, Acosta reported its use as an aphrodisiac in the East Indies, stating that "he who partakes of it is deprived of his reason for a long time, laughing or weeping or sleeping . . . at times he appears to be in his right mind, but really being out of it. . . ." In China the plant was often considered sacred, and it was believed that, when Buddha preached, heaven sprinkled the plant with drops of dew or rain. The epithet *Datura* was taken by Linnaeus from the vernacular name *dhatura* or *dutra* in India, where knowledge of the intoxicating effects of the plant go back to prehistory and where it is still valued for its narcotic properties. *Datura metel* is a common admixture with cannabis preparations in India where it is also often employed criminally to this day. Leaves of the white-flowered form (often considered a distinct species, *D. fastuosa*) are smoked with cannabis or tobacco in parts of Africa and Asia. *Datura ferox,* an Asiatic species now widely distributed in warm areas of both hemispheres, is likewise esteemed as a narcotic and medicine in sundry regions.[307]

Although there is disagreement as to where *D. stramonium* is native, evidence seems to indicate that it was indigenous to the New World. It is employed as a narcotic in both hemispheres. Even the Algonquin and other tribes of the eastern woodlands of North America may have used *jimson weed* or *thorn apple* as an hallucinogen. The toxic medicine *wysoccan,* taken in adolescent rites of these Indians, is thought to have had *D. stramonium* as its chief active ingredient. In these initiatory rituals, youths were confined for long periods, given ". . . no other substance but the infusion or decoction of some poisonous, intoxicating roots . . ." and "they became stark, staring mad, in which raving condition they were kept eighteen or twenty days." They were said to "unlive their former lives" and commence manhood by losing all memory of ever having been boys.[304,307]

The real centres for the hallucinogenic use of *Datura,* however, lie in the American Southwest and Mexico and in South America.[338,347,348,349]

In North America, the species of apparently greatest importance is *D. inoxia,* more commonly known by the binomial *D. meteloides.* Other species, however are valued as hallucinogens in the Southwest and Mexico: *D. discolor* and *D. wrightii.* All of these, but especially *D. inoxia,* are called *toloache* in Mexico. A very interesting Mexican species is *D. ceratocaula,* a rather fleshy plant with thickish forking stems that grows in marshes and shallow water. Its vernacular name *torna-loco* ("maddening plant") indicates its strong narcotic properties. The ancient Mexicans, who referred to it as "sister of ololiuqui," held it to be very sacred. Priests about to employ it as a holy medicine addressed it reverently before even taking it.[307] Several other Mexican species may likewise be employed, locally at least, as divinatory narcotics or medicines: *D. kymatocarpa* and *D. reburra,* but evidence of their use is not yet conclusive.[23]

In the American Southwest, many tribes have utilized *Datura* ceremonially, especially amongst Indians of California, Arizona and New Mexico. It has been said that *D. inoxia* was the most universally used drug by the natives of this region. The Zunis call *D. inoxia a-neg-la-kya* and value it as an hallucinogen, anaesthetic, and as a poultice for wounds and bruises. This plant belongs exclusively to the rain priests amongst the Zunis, and they are the only men who may collect the roots. When these priests commune at night with the feathered kingdom, they put powdered roots into the eyes; and they chew the roots to ask the spirits of the dead to intercede with the gods for rain. *Datura inoxia* is basic to adolescent initiation rites of many southwestern tribes. The Yumans take it to induce dreams and gain occult powers and predict the future. Many Indians believe that they can acquire supernatural helpers through the drug, and much secret knowledge is thought to be gained during the ceremony. Yokuts usually take the seeds only once during a lifetime, but boys who are studying to be shamen must undergo the intoxication once a year.[305,307]

The importance of *Datura* amongst Mexican Indians—both now and in pre-Conquest times—was even greater. One of the earliest accurate accounts is that of Hernández who listed many therapeutic uses of *toloatzin,* warning that excessive use of it could drive the patient to a madness characterized by "various and vain imaginations." The modern Tarahumares, for example, add *D. inoxia* to *tesquino* (a drink prepared from sprouted maize) to make it strong, and the roots, seeds and leaves of this species are the basis of a drink taken ceremonially to induce visions. Tarahumare medicine men take it to help them diagnose disease. Amongst some Mexican Indians, toloache is considered an hallucinogen inhabited by a malevolent spirit, unlike peyote.[109,307]

All of the indigenous South American species of *Datura*—arborescent—belong to the subgenus *Brugmansia.* Most of them have become of great horticultural interest. All are cultigens, unknown in a truly wild state, which bespeaks long association with man and his agricultural practices because of their medicinal

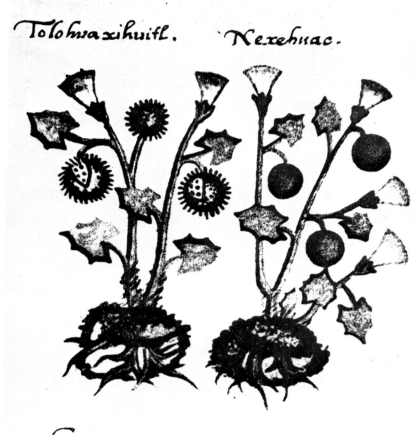

Tolohuaxihuitl.   Nexehuac.

Contra laterum dolorem.

Figure 54. *Datura inoxia* or tolohuaxihuitl of ancient Mexico. From E. W. Emmart (Ed.): *Badianus Manuscript* (1940).

and narcotic value in his primitive societies. They are all probably chromosomally aberrant and usually set viable fruit erratically. There are only a few species, but these species exhibit extreme variation: *D. arborea, D. aurea, D. candida, D. dolichocarpa, D. sanguinea* and *D. vulcanicola* in the Andean highlands from

Colombia south to Chile; and *D. suaveolens*, native to the warmer lowlands. The taxonomy of these tree species is (as is true of many cultigens) somewhat poorly understood. A recent taxonomic study has suggested that this subgenus should be treated as comprising three species—*D. candida, D. sanguinea* and *D. suaveolens* —and numerous cultivars. They are known by many local names, the most often encountered being *borrachero, huacacachu, huanto, chamico, campanilla, floripondio, maicoa, tonga, toa.*[24,306,314]

   Not only is there a suspicion that all tree-*Datura* species are

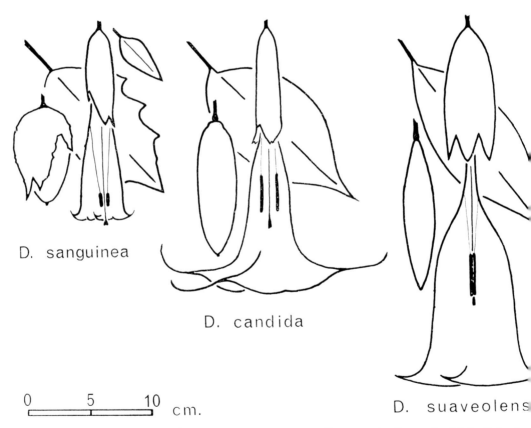

D. sanguinea

D. candida

0   5   10
cm.

D. suaveolens

Figure 55. Three types of flowers in *Datura*, section *Brugmonsia*. Drawn by M. L. Bristol. Courtesy Botanical Museum Leaflets, Harvard University.

cultigens, but these plants offer complex biological problems resulting from their close association over the millennia with man. Bristol has written: "Many writers have noticed the frequency with which tree-Daturas are associated with human habitations, but the extent of this association and its implications have not been fully understood. . . . The northern Andes . . . is the centre of variability and probable area of origin of this group." [48,49]

Datura suaveolens is recognized as toxic and narcotic in the western Amazon, where it is still employed hallucinogenically, both alone and as an admixture with other drugs such as *Banisteriopsis.* It is in the Andes, however, that the hallucinogenic use of the tree-*Datura* species is centered. Although the species are utilized widely, the literature is very deficient and has reported only a few of the many tribes that use them in their daily life. Scores of aboriginal peoples from Colombia south to Ecuador, Peru, Bolivia and Chile value these trees as ritualistic hallucinogens and medicines: Chibchas, Chocos, Inganos, Kamsás, Sionas, Kofans of Colombia; the Quechuas of Bolivia, Ecuador and Peru; the Mapuche-Huilliches of Chile; the Canelos, Piojos, Omaguas, Jivaros, Zaparos of eastern Ecuador. Some tribes along the Pacific coastal area of Colombia and Ecuador likewise utilize *D. suaveolens.* In Chile, the Mapuches value *D. candida* and *D. sanguinea* as a correctional medicine for unruly children; the Jivaro believe that the spirits of their ancestors speak to and admonish recalcitrant children during the hallucinations that Datura induces. The Chibchas of pre-Conquest Bogotá gave *chicha* (a fermented maize drink) with *Datura* (probably *D. aurea,* *D. candida* or *D. sanguinea*) to wives and slaves of dead warriors or chieftains to induce a state of stupour before they were buried alive to accompany their husbands and masters on the long trip. The Indians living north of Bogotá, at Sogamoza, used *D. sanguinea* in ceremonies at the Temple of the Sun. The narcotic prepared from this red-flowered species, known locally as *tonga,* is reputedly stronger than that from the white-flowered species. In Peru, although the acceptance of Christianity by many Indian groups has altered the ancient beliefs and customs, many natives still maintain that such plants as *D. sanguinea* permit them

Figure 56. *Datura suaveolens*. Mocoa, Putumayo, Columbia. Photograph by R. E. Schultes.

to communicate with ancestors or others in the spirit world. The Indians in the Peruvian town of Matucanas, for example, believe that *D. sanguinea* will reveal to them treasures preserved in ancient graves or huacas, from which comes the local name *huacacachu* or "grave plant." [285,307,348,349,413]

Throughout the Andes, excepting in southern Chile, the mode of preparation and use of *Datura* differed from tribe to tribe, but most often the drug was taken in the form of pulverized seeds dropped into fermented drinks or as an infusion of leaves and twigs. Intoxication is characterized usually by initial effects so violent that physical restraint must be practiced until the partaker passes into the stage of a deep sleep, during which hallucinations occur. The medicine man interprets the visions or visitations of

Figure 57. *Datura sanguinea*. Near Zipaquirá, Cundinamarca, Colombia. Photograph by R. E. Schultes.

the spirits that enable him to diagnose disease, apprehend thieves and prophesy the future of tribal affairs and aspirations.[349,418]

As in North America, there is usually no reliable information on which to base a definite identification of the species of *Datura* employed. References concerning their use are almost invariably found in the literature and are not based upon voucher specimens. Consequently, the species involved in each instance must usually be guessed from phytogeographical or ecological reasoning or, when available, from colour or other morphological descriptions; occasionally, the use of a vernacular name helps. However, since most, if not all, species of *Datura* contain similar tropane alkaloids varying only in relative concentrations, this method of discerning which species are used does not pose any serious problem or uncertainty, nor is there usually a danger of confusing the *Datura* with some other hallucinogen. The time is long overdue when comparative chemical studies of all species identified by voucher specimens be undertaken, for, if the taxonomy of this genus may be described as still uncertain (which is truly the case), our exact knowledge of the chemistry, from a comparative viewpoint, is chaotic; this is the result primarily of careless or superficial plant identifications and failure to file away an authenticating specimen for each analysis.[347]

The taxonomy and chemistry of the tree-*Datura* species is complicated in certain Andean regions by the prevalence of curious atrophied "races" of some of the species. These are valued by the natives because of the bizarre appearance, making them attractive in magic, or because of different physiological effects due, presumably, to varying chemical composition. These "races" are reproduced vegetatively by planting pieces of stem in damp soil, and represent virtually different clones with very definite native names. An excellent example of this biologically interesting phenomenon has been studied in the high, mountain-girt Valley of Sibundoy in southern Colombia, where the Kamsá and Ingano Indians employ several species—*D. candida, D. dolichocarpa, D. sanguinea*—and a number of clones of *D. candida*. They are so highly atrophied that they may possibly represent incipient "varieties" as the result of mutations. Some are such monstrosities that their botanical identification to known species has, until recently, defied efforts. They are said to differ in their narcotic effects and strength, being stronger or weaker than "normal" daturas. These "races" of Sibundoy have recently been studied by Bristol who has assigned cultivar names to them.[48,49,334,335,338,347,349]

There has as yet been no satisfactory explanation of why there should be such a concentration in this one high locality of so many atrophied forms. Obviously, the very deep interest of the indigenous population in the use of narcotics has helped in the preservation of these aberrant types which, had they not appealed to the Indian practitioner, would undoubtedly have become extinct. But

Figure 58. One of the atrophied "varieties" of *Datura candida*. Sibundoy, Putumayo, Colombia. Photograph by R. E. Schultes.

how did they arise? One suggestion attributes the condition to an extreme virus infection, not uncommon to the Solanaceae.[334,335] Another suggestion quotes Blakeslee's work that shows that at least the herbaceous daturas ". . . demonstrated a great range of variability and the spontaneous appearance of many unusual characteristics. Of the 541 gene mutations encountered, 72 appeared following heating, wounding and ageing or spontaneously in nature. Recessive genes controlling leaf shape, flower size, shape and colour, and fruit form are amongst those uncovered. It is entirely possible that these single recessive genes affecting taxonomically significant characters are present also in the tree-Daturas." [24]

It will undoubtedly be sufficient here to discuss in detail the chemistry of only one of the herbaceous species of *Dutura* and one of the arborescent members of the subgenus *Brugmansia*.

All species of *Datura* contain as their principal active component scopolamine (hyoscine). Furthermore, all are very similar

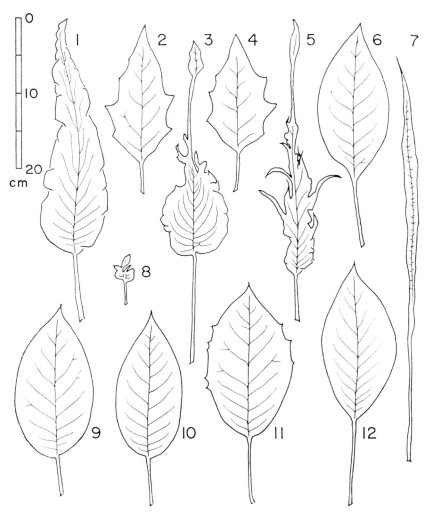

Figure 59. Atrophied "varieties" of the tree-daturas and of *Methysticodendron* (7). Sibundoy, Columbia. Drawn by M. L. Bristol. Courtesy *Botanical Museum Leaflets*, Harvard University.

with respect to their content of minor alkaloids. It will undoubtedly be sufficient, therefore, to discuss here in detail the chemistry of two representative herbaceous species—*D. metel* and *D. inoxia*, and one arborescent member of the subgenus *Brugmansia—D. candida*.

In the various parts of *D. metel*, the following total alkaloid contents were reported: fruits 0.12 percent, leaves 0.2 to 0.5 percent; roots 0.1 to 0.2 percent; seeds 0.2 to 0.5 percent. Together with the main component, scopolamine, minor alkaloids were found: meteloidine, hyoscyamine, norhyoscyamine, norscopolamine, in addition to two alkaloids not belonging to the tropane group: cuscohygrine and nicotine. Meteloidine, the first isolated from *D. meteloides* (=*D. inoxia*) is the ester of 6,7-β-dihydroxytropine with tiglic acid.

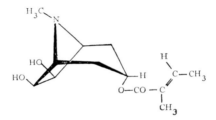

Meteloidine

The composition of the alkaloidal components of *D. inoxia* was found to differ only slightly in the minor components from that of *D. metel*.

Systematic investigations on the alkaloids of the arborescent species of *Datura* (subgenus *Brugmansia*) have been carried out only very recently. Evans and his co-workers published a review of their investigations and of that of other groups, designed to correlate chemical findings in these species with accurate taxonomic studies.[51]

The isolation of scopolamine as the major alkaloid, together with norscopolamine, atropine, meteloidine, oscine and noratropine from the aerial parts of *D. candida* and 3a, 6β-ditigloyloxytropan-7β-ol, scopolamine, norscopolamine, 3a-tigloyloxytropane, meteloidine, atropine, oscine, noratropine and tropine from

Figure 60. *Datura candida*. Sibundoy, Putumayo, Colombia. Photograph by
M. L. Bristol.

the roots, indicated that there is little qualitative difference
between the alkaloids of this plant and the other closely related
species thus far investigated. Furthermore, the same alkaloids are
found also in the various Sibundoy cultivars. The total alkaloid

content varies between 0.3–0.55 percent of the dry plant matter, with a percentage of scopolamine varying between 31 and 60 percent.

Descriptions of only two of the important species will be given: one Old World species of section *Dutra* (*D. metel*) and one species of section *Brugmansia* (*D. candida*).

Datura metel Linnaeus Sp. Pl. (1753) 179.

Spreading herb, sometimes becoming shrubby, 3–6 feet tall. Stems terete, glossy, glabrous, in age with persistent leaf scars. Leaves ovate or triangular-ovate, sinuate, nearly entire to deeply but distantly dentate, acute, 15–21 cm long, 7.5–10.5 cm wide; petiole up to 2 cm long. Flowers solitary, large, purple, white or yellow (sometimes purplish outside, whitish within); pedicels up to 1 cm long. Calyx regular, minutely pubescent, 5–7 cm long, with 5 short, triangular-lanceolate teeth. Corolla tubular, funnel- or trumpet-shaped, almost circular when expanded, usually with 5 equidistant radiating nerves terminating in short tails, simple, double or triple by petaloid outgrowth of stamens and inner corolla surfaces, 14–15 cm long. Style 11–13 cm long. Capsule globose, nodding, conspicuously tuberculate or muricate, 4–6 cm in diameter, borne on short, thick inclined peduncle.

This species, of which there are several rather distinct types, is indigenous to Asia but now ranges widely in tropical and subtropical Asia, Africa and America.

Datura candida (*Pers.*) *Safford*, Journ. Wash. Acad. Sci. 11 (1921) 182.

Tree 6–17 feet tall, tardily lignified, with some dichotomously forked branches. Leaves ovate, oblong-elliptic or ovate-lanceolate, entire or coarsely dentate (never angular), glabrous or slightly and minutely pubescent, usually paired with one smaller than other, up to 25 cm long, 15 cm wide, usually long-petiolate. Flowers very large and conspicuous, usually pendulous, 20 cm long or longer, usually trumpet-shaped, white, with pronounced musk-like aroma, especially after sundown. Calyx spathelike, long acuminate, 1.5–3 cm broad, 1- to 4-toothed. Corolla 20 cm long or longer, white, basal tubular part terete, very slender and enclosed wholly in calyx, limb flaring broadly with long (4–9 cm) recurved teeth, margin of limb between teeth entire or rounded. Anthers distinct, not conglomerate, yellow. Fruit green, oblong-cylindrical to fusiform, without persistent calyx, long-acuminate, long-pedunculate, usually 15–25 cm long. Seeds angular with thick, suberose testa.

This is the common white-flowered species of the Andean highlands and is extremely variable. It has been confused in the taxonomic literature with *D. arborea,* a species of much more limited and disjunct distribution in the highlands and with *D. suaveolens* of warmer, lower parts of South America. Almost always when the binomial "*D. arborea*" has been used in the botanical or horticultural and chemical literature, it refers to the concept now known as *D. candida.*

Datura candida was described from material collected in Peru, but it occurs spontaneously throughout most of the Andean ranges and has been widely introduced into cultivation in all warmer parts of both hemispheres.

A list of species known or suspected to be employed as hallucinogens follows:

Section *Stramonium*

Datura stramonium *Linnaeus* Sp. Pl. (1753) 179.
Section *Dutra*
  Datura inoxia *Miller* Gard. Dict., ed. 8 (1768) unpag.
    (**Datura meteloides** *DeCandolle ex Dunal,* DeCandolle Prodr. 13, 1 (1852) 544)
  Datura discolor *Bernhardi* Trommsdorf N. Journ. Pharm. 26 (1833) 149.
  Datura metel *Linnaeus* Sp. Pl. (1753) 179.
    (**Datura fastuosa** *Linnaeus* Syst. Nat., ed. 10 (1759) 932.)
  Datura wrightii *Regel* Gartenfl. 8 (1859) 193.
Section *Ceratocaulis*
  Datura ceratocaula *Ortega* Decas. Primas 1 (1971) 11.
Section *Brugmansia*
  Datura arborea *Linnaeus* Sp. Pl. (1753) 179.

Datura aurea Lagerheim Gartenfl. 42 (1893) 33.

Datura candida (*Pers.*) *Safford*, Journ. Wash. Acad. Sci. 11 (1921) 182.

Datura dolichocarpa (*Lagerh.*) *Safford*, Journ. Wash. Acad. Sci. 11 (1921) 186.

Datura sanguinea Ruíz *&* Pavón Fl. Peruv. 2 (1799) 15.

Datura suaveolens Humboldt *&* Bonpland *ex* Willdenow Enum. Hort. Berol. (1809) 227.

Datura vulcanicola Barclay, Bot. Mus. Leafl., Harvard Univ. 18 (1959) 260.

Hyoscyamus Linnaeus

A genus of some 20 species of Europe, northern Africa, south-western and central Asia.

Hyoscyamus niger, an annual or biennial native to Europe but now naturalized and growing spontaneously across north-temperate Asia and North America, is commonly called *henbane* in English because of its extreme toxicity. It has been employed medicinally since very early times. It has been valued as a sedative and anodyne for inducing sleep, both the leaves and seeds finding use in pharmacy; however, hallucinations often accompany its incautious ingestion. Accidental poisonings are recorded from medieval times and earlier in Europe, and it is thought to have been used as one of the active ingredients in witches' brews and other toxic preparations of the Dark Ages, when visual hallucinations and flights of fancy were effects frequently sought by those practising this form of witchcraft. Other species, especially *H. muticus*, may likewise have similarly been employed as narcotics.[217,418]

The principal alkaloid of *H. niger* is hyoscyamine, isolated for the first time from the seeds of this plant by Geiger and Hesse in 1833.[113a] But an essential proportion of the total alkaloids, up to 50 percent of hyoscyamine, consists of scopolamine. Minor alkaloids of *H. niger* were isolated: tropine, scopine, apoatropine and cuscohygrine. The total alkaloid content of the various parts of the plant was found to be as follows: leaves 0.04 to 0.08 percent; roots 0.16 percent; seeds 0.06 to 0.1 percent.[149a]

Hyoscyamus muticus shows essentially the same alkaloidal composition as *H. niger*, but the total alkaloid content tends to be higher: leaves 1.4 percent; seeds 0.9 to 1.34 percent.

Hyoscyamus niger Linnaeus Sp. Pl. (1753) 179.

Annual or biennial with rank odour, glandular-pubescent, usually up to 3½ feet tall. Stem stout, basally woody. Leaves oblong-ovate, nearly entire or with a few large teeth, lower ones up to 15–20 cm long, petiolate; cauline ones smaller, amplexicaul. Flowers 6–8, subsessile in scorpioid cymes, about 2 cm; bracts foliose. Calyx urceolate, 5-fid; teeth triangular, acuminate, growing to enclose fruit and becoming rigid with sharp point at maturity. Corolla funnel-shaped, rather zygomorphic, often laterally split, 2–3 cm in diameter, yellow with purplish veins; lobes rounded. Stamens inserted at base of corolla tube; anthers purple. Style filiform. Stigma capitate. Capsule smooth, many-seeded, opening by apical lid.

Hyoscyamus niger is native to Europe, western Asia and northern Africa.

Latua Philippi

A monotypic genus endemic to the coastal mountains of central Chile, apparently nowhere abundant.

The unusual and rare spiny shrub or small tree *Latua pubiflora* was formerly employed by medicine men, especially in the province of Valdivia, as a virulently toxic agent capable of producing delirium, hallucinations and, on occasion, permanent insanity. The plant was locally called *latué* or *árbol de los brujos* ("sorcerers' tree"). No cult or ritual apparently attended its use, but the plant was widely respected and feared in the region. It was said that a madness of any duration could be induced at the will of the medicine man, according to the strength of the dose. The dosages were a closely guarded secret amongst the native practitioners. Accidental poisonings also happened, since *L. pubiflora* closely resembles *Flotowia diacanthoides,* a shrub known as *tayu* and employed in folk medicine in a decoction of the bark to treat bruises and blows. The two plants were often confused.[243,271,345,347,349]

This same species-concept has been known under the name *L. venenosa* Philippi.

A chemical investigation of the leaves of this plant revealed the presence of 0.18 percent hyoscyamine and 0.08 percent scopolamine. The dried fruits did not contain any alkaloids, whereas fresh fruits are said by the aborigines to be highly toxic.[40a]

Latua pubiflora (Griseb.) Baillon Hist. Plant. 9 (1888) 334.

Shrub to small tree, 9–30 feet in height. Bark thin, streaked with corky, longitudinal fissures, becoming somewhat rough, reddish to greyish brown. Branches smooth, armed with spines. Branchlets of current year's growth covered with a yellowish brown pubescence, becoming glabrous. Spines erect, arising as modified branches in the leaf axils, rigid, 20 mm long, usually with small basal leaf and one or two minute cataphylls towards tip. Leaves alternate, fascicled on short shoots or scattered on long shoots, narrowly elliptic to oblong-lanceolate, acute, entire to erose-serrate, basally attenuate, 3.5–12.0 cm long, 1.5–4.0 cm wide, pilose, but glabrescent with age; petiole usually short, 2(15)mm, pilose, glabrescent. Peduncle solitary, arising from axil of spine and basal leaf, erect, one-flowered, 5–9(20)mm long, tomentose, basally scaley; scales ciliate, ovate, about 2 mm long. Calyx 5-parted, campanulate, persistent, 8–10 mm long (in fruit, 11–16 mm long), tomentose, pale green to purplish; lobes valvate acute, triangular, about 3 mm long. Corolla larger than calyx, 5-parted, urceolate, basally narrowed, slightly constricted below the limb, inflated, 3.5–4.0 cm long, 1.5 cm in diameter at the middle, densely pilose without, magenta to red-violet; lobes induplicate-valvate, short, trilobate, recurved, about 5 mm long. Stamens 5, inserted basally on corolla; filaments of differing lengths, slightly exceeding corolla, filiform, 3–4 cm long, adnate for 8 mm, basally pilose, glabrous above, pink; anthers bilocular, longitudinally dehiscent, elliptical, 2 mm long, brownish. Ovary ovoid, basally gibbous, bilocular; style filiform, equaling corolla, pink; stigma short, semicircular, slightly bilobate, green. Fruit fleshy berry, globose, bilocular, base of style persistent, 2 cm in

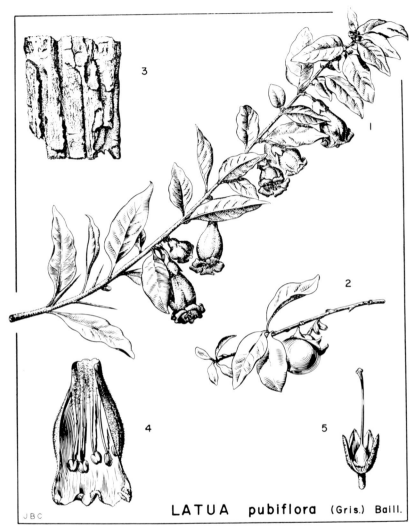

Figure 61. Drawn by J. B. Clark.

diameter, pale green to yellow. Seeds numerous, somewhat reniform or irregularly shaped, usually flattened ventrally, with thick,
reticulate-cellular, dark brown to black testa.

Native to the coastal mountains of central Chile between
Valdivia and Chiloé, *L. pubiflora* is found in humid mountain

forests or adventive in fields and pastures between 1500 and 2000 feet altitude.

Mandragora Linnaeus

There are six species of *Mandragora* known, native to the region from the Mediterranean eastward to the Himalayas.

Mandragora officinarum, the famous *mandrake* of Europe, has long been known for its toxic properties and for its actual and presumed medicinal virtues. Its complex history in folklore as a magic plant cannot be equalled by any other species. In folk medicine, mandrake was a panacea, recommended as a sedative and hypnotic agent in treating nervous conditions and pain, its toxicity notwithstanding. The fear and respect in which Europeans of the Middle Ages and earlier held the mandrake and many of its folk uses were inextricably bound up with the so-called "Doctrine of Signatures." Roots of *Mandragora,* in the shape of a little man or woman, were named *Alraune, Erdmännchen* or *Erdweibchen* and were highly praised as amulets and used for all kinds of magic purposes in the Middle Ages. One of its real practical uses during the Middle Ages was that of a pain-killer during surgery. But it was further employed for many other illnesses and abnormal conditions. In many regions, it enjoyed a reputation as an aphrodisiac.[217,418]

Like *Atropa* and *Hyoscyamus,* there seems to be no reason to doubt that *Mandragora* served as one of the potently hallucinogenic constituents common in witches' brews during the Dark Ages and early Middle Ages in Europe. It may be one of the reasons for the present German name of the plant: *Hexenkraut.*[418]

The first reliable chemical investigation of the alkaloids of *Mandragora* was carried out by Hesse in 1901.[150b] Hyoscyamine was found to be the major component, with small quantities of scopolamine, pseudo-hyoscyamine (nor-hyoscyamine) and a new alkaloid which was named mandragorine. A recent analysis of the roots of *Mandragora,* with the use of modern isolation and identification procedures, yielded supplementary results.[382a] The total alkaloid content was found to be 0.4 percent, the principal alkaloid being hyoscyamine, followed by scopolamine and atropine in the ratio of 18:2.5:1. Cuscohygrine, which Phokas[269a] showed to be identical with Hesse's mandragorine, was also isolated in small quantities.

$$CH_2COCH_2$$

Cuscohygrine
(Mandragorine)

Mandragora officinarum Linnaeus Sp. Pl. (1753) 181.

Stemless herb with fleshy root more or less branched (fancifully resembling human form). Leaves mostly radical, wrinkled, crisp, ovate to ovate-oblong, entire or somewhat sinuate-dentate, lower obtuse, uppermost often acute or acuminate, glabrous or minutely pubescent, 15–25 cm long (becoming longer—up to 40 cm—in fruiting plants). Flowers short-pedunculate, solitary, pale purple or white; peduncles as long as or longer than calyx. Calyx foliose, 5-fid, continuing to grow in fruit; lobes lanceolate. Corolla campanulate, 5-fid half its length, lobes broad, overlapping, sinuses induplicate between lobes, bluish or whitish green. Stamens 5, inserted below middle of corolla tube; filaments threadlike; anther cells almost parallel. Ovary implanted on disk. Style long, filiform. Stigma capitate, more or less 2-lobed. Berry fragrant, succulent, globose, yellow, 3 mm in diameter.

Mandragora officinarum is indigenous to the Mediterranean areas of Europe.

Methysticodendron R. E. Schultes

An anomalous, monotypic genus (representing possibly an aberrant form of a tree-*Datura*) known only from cultivated trees in the Valley of Sibundoy in the southern Andes of Colombia.

In addition to the several species of tree-daturas and the clones or "races" of *D. candida*, the Indians of Sibundoy in Colombia cultivate an extraordinary and enigmatic tree which has been described as a new genus, *Methysticodendron amesianum*, locally known as *culebra borrachera* or, in the Kamsá language, as *mits-*

Known only from cultivated material from the Valley of Sibundoy, Putumayo, Colombia, at about 7,500 feet altitude.

Methysticodendron differs from the tree-*Datura* (section *Brugmansia*) by its very deeply lobose corolla; the narrowly ligulate leaves; the three conduplicate carpels with three free styles (with undivided stigmatic areas) that retain the conduplicate condition; the concave structure of the styles that leaves the ovule cavity somewhat open at the apex, exposing the apical ovules, until, about one third of the way down, the carpel walls fuse to form a trilocular syncarp.[328]

It has been suggested that this concept represents a species of *Datura* that has been highly atrophied as a result of viral infection, but no definite proof of this suggestion has been forthcoming. Another explanation held that it might have resulted from ". . . the action of a single pleiotropic gene mutation . . . a monstrosity of some *Datura* species of subg. *Brugmansia*. . . ."[48,49] It has further been considered to represent a cultivar of *D. candida* and has been named as follows:

Datura candida (*Pers.*) *Safford* cv. **culebra** Bristol in
Bot. Mus. Leafl., Harvard Univ. 22 (1969) 218.

ACANTHACEAE

Justicia Linnaeus

A genus of more than 350 species, many of them aromatic, of the tropical and subtropical parts of both hemispheres.

Recent ethnobotanical studies have revealed the use of the pulverized leaves of *Justicia pectoralis* var. *stenophylla* as an occasional additive in the hallucinogenic snuff prepared basically from the bark-resin of *Virola* in the Venezuelan headwaters of the Orinoco and along the northern affluents of the Rio Negro of Brazil.

The Waiká and other Indians of this general area are accustomed on occasion to add the aromatic *Justicia* powder "to make the snuff smell better." Many localities have *J. pectoralis* var. *stenophylla* under cultivation in dooryards for this purpose. There is evidence that the natives in some of the Orinoco localities may prepare an hallucinogenic snuff exclusively from *Justicia* and that other species of the genus may also be involved.[62,340]

Chemical examination of *Justicia* is still incipient. Preliminary indications of the presence in leaves of *J. pectoralis* var. *stenophylla* of N,N-dimethyltryptamine still lack corroboration with additional material.[347,350] If this base is unquestionably found in *Justicia*, it will be the first time that tryptamines—or, in fact, any hallucinogenic principle—have been reported for the Acanthaceae.

Justicia pectoralis Jacquin var. *stenophylla Leonard,* Contrib. U. S. Nat. Herb. 31 (1958) 615.

Herb up to 1 foot in height, compact. Stems erect or ascending, sometimes rooting at lower nodes, subquadrangular, somewhat grooved, glabrous or occasionally hirtellous; internodes short, usually less than 2 cm long. Leaf blades numerous, very narrowly

Figure 65. Waiká Indian with leaves of *Justicia pectoralis* var. *stenophylla* for use as admixture in *Virola*-snuff nyakwana. Rio Tototobí, Amazonian Brazil. Photograph by R. E. Schultes.

lanceolate, 2–6 cm long, 1–5 cm wide, acuminate, basally cuneate, entire, glabrous above and below, petioles slender, up to 6 mm long. Inflorescence often becoming dense, up to 10 cm long but usually much shorter, with glandular and eglandular hairs. Flow-

Figure 66

ers small: calyx 5-fid, segments subulate, about 2 mm long, 0.25 mm wide, puberulous; corolla white or violet, sometimes purple-spotted, about 7–8 mm long, externally slightly pubescent; stamens exserted about 1 mm beyond corolla-throat; style about 7 mm long. Capsules clavate, about 8 mm long. Seed flattish, about 15 mm broad, rough, red-brown.

This variety differs from *J. pectoralis* mainly in its smaller stature and in having very narrowly lanceolate leaves and a shorter inflorescence. While *J. pectoralis* is widespread in the West Indies and continental tropical America, the variety is known only from eastern Colombia and the adjacent parts of Amazonian Brazil where it is often semi-cultivated in yards and grows spontaneously in open spaces.

RUBIACEAE

The Rubiaceae is a natural family in the order Rubiales. One of the largest of phanerogamic families, it comprises some 500 genera and more than 6500 species, mainly tropical trees, shrubs and herbs. The family is divided into usually three sections: Cinchonoïdeae, with eight tribes; Rubioïdeae, with eleven tribes; Guettardoïdeae, with one tribe. A few genera are of economic importance as the source of caffeine beverages (*Coffea*), drugs (*Cephaelis, Cinchona*) and dyes (*Rubia*).

Psychotria Linnaeus

Psychotria belongs to the Rubioïdeae and contains more than 700 species, mostly small trees of the warmer parts of both hemispheres.

Since the Rubiaceae represents a very large family and one that is rich in alkaloids, it is curious that only the genus *Psychotria* has been employed for hallucinogenic purposes. Furthermore, this use has been discovered only in the past three or four years. It is even more interesting that *Psychotria* appears to be utilized only as an admixture with other hallucinogenic plants.

In several far-separated Amazonian localities, leaves of *Psychotria* are, on certain occasions, added to the psychotomimetic beverage made from *Banisteriopsis caapi* or *B. inebrians*. The Kofan Indians of Amazonian Colombia and Ecuador add the leaves of *P. viridis* (reported, as a result of misidentification, as *P. psychotriaefolia*) to lengthen and strengthen the visions induced. The Kashinahua of eastern Peru and western Brazil employ the leaves of two different species of *Psychotria*, known by the native terms *nai-kawa* and *matsi-kawa*. The identity of these two species is uncertain, because only sterile material could be collected; but the nai-kawa has been referred to one of the following species: *P. horizontalis*, *P. catharginensis*, *P. marginata* or *P. alba*, all of which grow abundantly in the region. Natives at Tarauacá in the Acre of Brazil, where the use of the hallucinogenic ayahuasca is employed by both Indians and Brazilians and where it is renowned for its potency, prepare the drink from *Banisteriopsis* with leaves of *P. viridis* added.[78,278,279,340,347]

The chemical composition of the *Psychotria* leaves explains why this plant is utilized as an additive and not as the sole base of a drink. They contain N,N-dimethyltryptamine[78] which is inactive when taken orally without a monoamine oxidase inhibitor. When the leaves are mixed with material from *Banisteriopsis* that contains β-carboline alkaloids, which are monoamine oxidase inhibitors, the tryptamine is effective in an orally administered form and in reality does materially alter the intoxication.[78]

Psychotria viridis Ruíz & Pavón, Fl. Peruv. 2 (1799) 61, t.210, fig. b.

Shrub or small tree, up to 14 feet in height, glabrous throughout. Stipules large, acuminate, thin, brownish, caducous. Leaves short-petiolate, obovate or obovate-oblong, acute or short-acuminate, basally long-cuneate, 8–15 cm long, 2.5–5 cm wide. Inflorescence terminal, pedunculate, spicate-paniculate, shorter than leaves, up to 10 cm long, lower branches more or less verticillate. Flowers sessile in distant glomerules, very small, usually 4 mm long; corolla greenish white, not basally gibbous. Fruit small, drupaceous.

Ranging in forests throughout the Amazon basin north to Central America and Cuba.

COMPOSITAE

The Compositae, belonging to the order Campanulatae and one of the largest families of spermatophytes, comprises 900 to

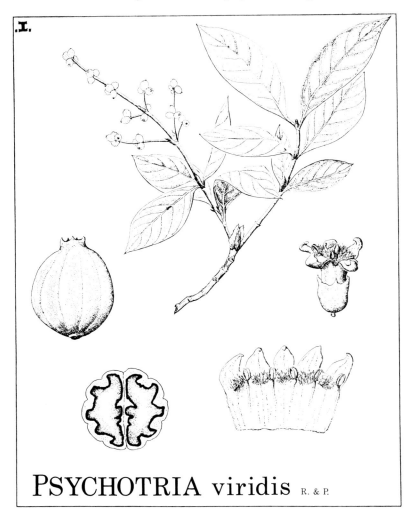

PSYCHOTRIA viridis R. & P.

Figure 67. Drawn by I. Brady.

1000 genera and 15,000 to 20,000 species. It represents probably the evolutionarily most advanced dicotyledonary family. Nearly cosmopolitan and living in almost every ecological situation, being rarest in tropical rain forests, most composites are herbaceous, but a few are trees and shrubs. Related obviously to the Brunoniaceae, Campanulaceae, Goodeniaceae and Stylidiaceae, the family is highly natural and is well marked by its characters, the most striking of which is the massing together of the flowers into dense heads subtended by an involucre of several to many bracts that often persist into the fruiting condition. It is divided usually into two sections: Tubiflorae, with about 11 tribes; and the Liguliflorae, with a single tribe. In relation to the size of the family, there are relatively few species of economic importance.

Calea Linnaeus

The genus *Calea* comprises about 100 species of the American tropics, growing especially in open savannah-like vegetation and on burned-over hillsides.

One of the most recently reported hallucinogens is *C. zaca-techichi*, an inconspicuous shrub ranging from central Mexico to Costa Rica. The specific name is derived from the Aztec word signifying "bitter grass," referring to the intense bitterness of this plant. It has been employed medicinally in folk medicine from earliest times, especially in the treatment of intermittent fevers, as an aperitif and as an astringent in cases of diarrhoea. While a decoction of the plant apparently does help reduce certain fevers, it is of no value in treating malaria, notwithstanding its reputation as an antimalarial. *C. zacatechichi* and other species have been utilized locally as insecticides.

Although there appears to be no evidence of a magico-religious cult in connection with *C. zacatechichi*, MacDougall has recently reported that the Chontal Indians of Oaxaca, who "believe in visions seen in dreams," employ this sacred plant to induce hallucinations. Crushed dried leaves are infused in water, and the resulting tea is imbibed slowly, after which the native lies down in a quiet place and smokes a cigarette of the dried leaves of the same plant. The Indian knows that he has taken a large enough dose when a sense of repose and drowsiness is experienced and when he hears his own heart and pulse beats. The Chontal medicine men, who assert that this plant is capable of "clarifying the senses," call *C. zacatechichi thle-pelakano* or "leaf of god." [229]

Further ethnobotanical studies of this interesting hallucinogen are indicated. Chemical studies that have been carried out in the past are of dubious value; more thorough investigations with

CALEA ZACATECHICHI Schlecht.

Figure 68. Drawn by I. Brady.

modern techniques are needed. A new alkaloid of undetermined structure has recently been isolated from *C. zacatechichi.*[347]

> ***Calea zacatechichi*** *Schlechtendal,* Linnaea 9 (1834)
>   589.

Freely branching shrub, pubescent to subglabrous. Leaves opposite, short-petiolate, ovate to triangular-ovate, marginally serrate, acute or acuminate, 2–6.5 cm long, rugose, nether surface more or less puberulent. Inflorescences small, dense, about 12-flowered umbellate cymes, pedicels usually shorter than heads, involucre several-seriate, phyllaries dry, without spreading tips, rays absent. Akenes more or less angled; pappus shorter than akenes.

Known from southern Mexico to Costa Rica, where it grows on scrubby hillsides and on disturbed land.

Chapter V

PLANTS OF POSSIBLE OR SUSPECTED HALLUCINOGENIC USE

Monocotyledonae
ARACEAE

THE ARACEAE, a natural family of the order Spathiflorae, and comprising about 115 genera and more than 2000 species, is divided into eight sections. A few species are temperate but most are natives of the tropics of both hemispheres—large and small herbs, climbing shrubs, epiphytic forms, marsh dwellers and even one wholly aquatic genus. A number of species contain a caustic or toxic latex; some produce large starchy edible rhizomes. Many are ornamentals. The aroids represent taxonomically and phytochemically one of the most poorly understood monocotyledonary families.

Acorus Linnaeus

A genus of two species of the north temperate and warmer parts of both hemispheres.

The Cree Indians of northern Canada chew the roots of *Acorus calamus,* known as *flag root, rat root* or *sweet calomel,* for various medicinal purposes and as a strong stimulant. Hoffer and Osmond [154] report that, on long walks, Indians dispel fatigue by chewing it and that a native informant reported that he felt as though he were walking a foot off the ground. Experiments with sophisticated subjects indicated that in large doses the root induces an experience similar to that of LSD. These investigators suggest that the active principles of flag root are asarone and β-asarone. Asarone shows a structural resemblance to mescaline, but their pharmacological activities are somewhat opposite; psychotomimetic properties have never been associated with asarone.[28,417]

Further ethnobotanical research must be carried out before it can definitely be asserted that North American Indians employ this plant hallucinogenically, and chemical and pharmacological investigation must then discover the principles responsible for such use.

Acorus calamus Linnaeus Sp. Pl. (1753) 324.

Homalomena Schott

Some 140 species of *Homalomena,* native to tropical Asia and South America, are known.

Natives of Papua are reported to eat the leaves of *ereriba,* a species of *Homalomena,* together with the leaves and bark of *Galbulimima belgraviana,* as a narcotic. The effects are a violent and crazed condition leading to deep sleep during which the partakers see and dream about the men or animals that they are supposed to kill. It is not yet clear what, if any, hallucinogenic principle may be present in this aroid.[25,347,349]

AMARYLLIDACEAE

The Amaryllidaceae is a large family of the order Liliiflorae and is related to the Liliaceae. Comprising 86 genera and more than 1300 species, it is divided into four subfamilies, several of which are separated as distinct families by some systematists. The family is distributed mainly in the tropics and subtropics and many members are adapted to xerophytic habitats. The amaryllidaceous ornamentals are numerous and curious, and many species find diverse uses in primitive societies and in modern agriculture and plant industry.

Pancratium Linnaeus

A genus of some 15 species, *Pancratium* is native mainly of Asia and Africa.

The Bushmen of Dobe, Botswana, are reported to employ *P. trianthum,* a bulbous perennial known locally as *kwashi,* as an hallucinogen. When the bulb is rubbed on an incision made on the head of a tribesman, visual hallucinations are said to be induced. Nothing more of this curious custom is known, and no studies towards the identification of a psychotomimetic constituent have been made. In west tropical Africa, *P. trianthum,* reputedly very toxic, is commonly planted in gardens around shrines.[345,347,349]

The genus *Pancratium* does possess powerfully toxic principles, including alkaloids. Some species find use in folk medicines as emetics; several are cardiac poisons; and one is said to have caused death through paralysis of the central nervous system.[347]

Pancratium trianthum *Herbert,* Ann. Nat. Hist., ser. 1 4 (1840) 28.

ZINGIBERACEAE

The Zingiberaceae is a family of perennial herbs, mostly aromatic, in the order Scitamineae. Estimates of the size of the family vary greatly—from 42 to 47 genera and from 400 to 1400 species of the tropics and subtropics of both hemispheres, mainly Indomalaysian. The family is divided into two subfamilies. Economically, the Zingiberaceae are important for their essential oils and as the source of a number of ornamentals.

Kaempferia Linnaeus

This genus has about 70 species of tropical Africa and from India to southern China and western Malaysia.

Rather vague reports indicate that *Kaempferia galanga* may be used hallucinogenically by natives in several parts of New Guinea where it is known as *maraba.*[26]

The rhizome of *galanga,* rich in essential oils, is highly prized as a condiment and in folk medicine in tropical Asia. In the Philippines, the rhizome, mixed with oils, is employed as a cicatrizant and is applied to boils and furuncles to bring them to a head. Other species—also condiments—are valued medicinally to hasten the healing of wounds and burns.[347,349]

Phytochemical corroboration of any hallucinogenic or other psychoactive properties of *K. galanga* is wholly lacking, but few if any investigations have been carried out on this species.

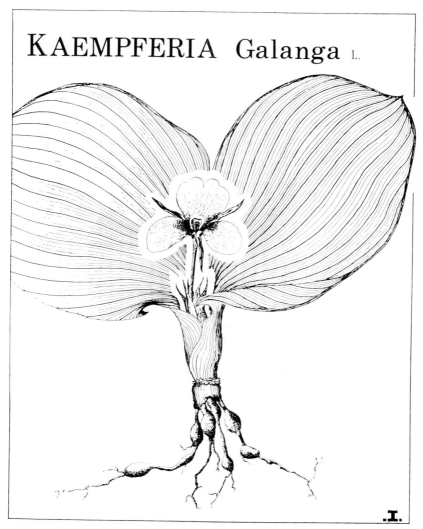

KAEMPFERIA Galanga L.

Figure 69. Drawn by I. Brady.

Kaempferia galanga Linnaeus Sp. Pl. (1753) 3.

Dicotyledonae
MORACEAE

The Moraceae, a member of the order Urticales and closely allied to the Urticaceae and Cannabaceae, comprises about 70 genera and well over 1000 mostly tropical species. The family is divided usually into four subfamilies or tribes, one of which—Cannaboideae—is now most frequently set aside as a separate family. The Moraceae are economically important as a source of edible fruits and as the source of fibre and ornamental plants.

Maquira Aublet

A small genus of large trees of tropical America.

The Indians of the Pariana region of the central part of the Brazilian Amazon formerly utilized an hallucinogenic snuff in ceremonies and dances. Its use has apparently died out with acculturation of the natives, but the source of the snuff, now known simply by its Portuguese name *rapé dos indios,* is still known to the general public of the area. It is said that the fruits of a very tall tree, *Maquira sclerophylla,* are the source of the snuff, but further ethnobotanical corroboration may be necessary to clarify all aspects of this curious narcotic.[330,334,335,338,345]

No chemical studies of this plant or of the snuff prepared from it have apparently been reported, and direct observation of the preparation and use of the narcotic powder have, as yet, not been possible.[347]

> ***Maquira sclerophylla*** *(Ducke) C. C. Berg,* Acta Bot. Neerl. 18 (1969) 463.
> ***Olmedioperebea sclerophylla*** Ducke, Arch. Jard Bot. Rio Jan. 3 (1922) 34.

This species has but recently been transferred into *Maquira.* It was described first as *Olmedioperebea sclerophylla,* in a genus of two species of jungle trees of the Brazilian Amazon. *Maquira sclerophylla* is known only from the central part of the Amazon Valley. This presumed hallucinogen has hitherto been referred to in the literature under its older name: *O. sclerophylla.*

AIZOACEAE

The Aizoaceae belongs to the order Centrospermae and comprises xerophytic herbs or undershrubs, chiefly South African but also of tropical Africa, Asia, Australia, California and South America. Estimates of the size of the family vary appreciably

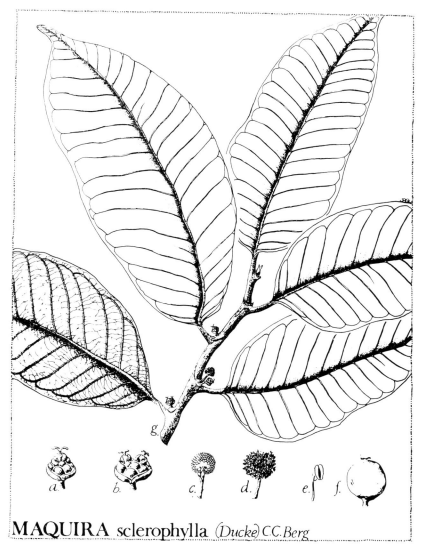

MAQUIRA sclerophylla *(Ducke)* C.C.Berg

Figure 70. Drawn by J. Gronim.

because of recent splitting of generic and specific concepts; estimates run from 80 to 130 genera, with some 1200 species being recognized. The family is divided into two subfamilies. Aizoaceous plants are horticulturally very important.

Mesembryanthemum Linnaeus

There are about 1000 species of *Mesembryanthemum* (sensu lat.); about two dozen South African species have been split off as representing a distinct genus, *Sceletium*.

The Hottentots of southern Africa, it was reported more than 225 years ago, chewed the root of a plant called *kanna* or *channa* to induce visual hallucinations. The masticated material was kept in the mouth for some time, after which "their animal spirits were awakened, their eyes sparkled and their faces manifested laughter and gaiety. Thousands of delightsome ideas appeared, and a pleasant jollity which enabled them to be amused by simple jests. By taking the substance to excess, they lost consciousness and fell into a terrible delirium." [218]

The identity of kanna, the narcotic, has remained a tantalizing mystery. This vernacular name is applied at the present time in South Africa to certain species of *Mesembryanthemum* (or *Sceletium*), especially to *M. expansum* and *M. tortuosum*. The roots and leaves of these two species are chewed and smoked in the hinterland areas but apparently not for hallucinogenic purposes. Several dozen species of *Mesembryanthemum* are known to contain alkaloids. [347]

The content of alkaloids in kanna ranges from 1.0 to 1.5 per cent. Mesembrine and mesembrenine are the main components and were isolated in yields of 0.7 per cent and 0.2 per cent, respectively. Half a dozen other alkaloids occur in very small quantities. [276]

Mesembrine   Mesembrenine

Mesembrine possesses sedative and cocaine-like properties and produces torpour in man.

There is, unfortunately, no direct line of evidence connecting the Hottentots and this curious custom—modern or past—with *Mesembryanthemum*. Lewin doubted that these aizoaceous plants could be responsible for the effects described. He suggested that the narcotic might have been *Cannabis sativa,* which the Hottentots do use habitually. He further pointed out that there were other intoxicating plants in South Africa, such as the anacardiaceous *Sclerocarya caffra* and S. *schweinfurthiana* which should be considered as candidates.

> *Mesembryanthemum expansum Linnaeus* Syst. Ed. 10 (1759) 1059.
>
> *Mesembryanthemum tortuosum Linnaeus* Sp. Pl. (1753) 487.
>
> *Sclerocarya caffra Sonder,* Linnaea 23 (1850) 26.
>
> *Sclerocarya schweinfurthiana Schinz,* Verh. Bot. Ver. Brand. Abhand. 29 (1888) 63.

HIMANTANDRACEAE

The Himantandraceae, belonging to the Ranales and related to the Magnoliaceae and Degeneriaceae, is a monogeneric, relict family of one or several species of eastern Malaysia and Australia

Galbulimima F. M. Bailey

The genus *Galbulimima* comprises from one to four species of large trees, the number of species varying according to the botanical authority consulted.

In Papua, the leaves and bark of *G. belgraviana* are taken with the leaves of a species of *Homalomena* to induce a violent intoxication that progresses into a sleep in which visions and dreams are experienced.[25,347,349]

The leaves and wood contain only minor amounts of basic material which has not been examined, but 28 alkaloids have so far been isolated from the bark. At the time of the first isolation of the alkaloids, it was thought that *Himantandra* was the correct name of the genus, and, consequently, the names bestowed on the alkaloids began with the syllable *him*, such as himbacine, himbosine, himgaline, etc. The galbulimima alkaloids are polycyclic piperidine derivatives. The structure of most of them has been elucidated. Himbacine showed antispasmodic activity with low toxicity, but there are no indications of hallucinogenic effects of these alkaloids.[295]

> ***Galbulimima belgraviana*** (*F. Muell.*) *Sprague* Journ.
> Bot. 60 (1922) 138.

GOMORTEGACEAE

The Gomortegaceae is a ranalean family of one genus, *Gomortega*, a very strict endemic of the southernmost Andes, related most closely to the Atherospermataceae and the Lauraceae.

Gomortega Ruíz & Pavón

A genus of one species, *Gomortega keule*, a large tree with essential oil in the leaves. It is said to occur in an area of only about 100 square miles.

The Mapuche Indians of Chile, who know *G. keule* as *keule* or *hualhual*, are said formerly to have valued this species as a narcotic. The intoxicating effects may possibly have been hallucinogenic. It is known that the fruits, especially in the fresh state, are inebriating, due possibly to their high concentration of essential oils.[347]

Chemical studies have apparently not been carried out on this anomalous species.[347]

> ***Gomortega keule*** (*Mol.*) *I. M. Johnston*, Contrib. Gray
> Herb., n.s., 3, no. 70 (1924) 92.

LEGUMINOSAE

Erythrina Linnaeus

A genus belonging to the Papilionoideae, *Erythrina* comprises about 100 tropical and subtropical species of both hemispheres. A number of species are highly toxic.

The reddish beans of several species of *Erythrina* may have

Figure 71. *Gomortega keule:* branch in fruit. Chile. Photograph by J. W. Walker.

been valued as hallucinogens in Mexico. Resembling seeds of *Sophora secundiflora*, they are frequently sold in modern Mexican herb markets under the name *colorines*. Further field studies in Mexico are necessary before an evaluation of this possible hallucinogenic use may be made.[347,349]

Sundry species of *Erythrina* contain alkaloids, all possessing the same tetracyclic ring system named erythran.[151] These alkaloids elicit a remarkable curariform activity,[80] but there is no pharmacological evidence of hallucinogenic activity.

Rhynchosia Loureiro

Related to *Erythrina* and also a member of the Papilionoideae, *Rhynchosia* is a tropical and subtropical genus of both hemispheres (but especially developed in America and Africa) of some 300 species.

The toxic principle of *R. pyramidalis* is unknown. A preliminary chemical and pharmacological examination of the seeds of *R. phaseoloides* was carried out by Santesson. The occurrence of an

alkaloid or gluco-alkaloid was presumed, based on positive alka-
loid- and glucoside-reactions, but without isolation of such a
compound. Crude extracts produce a kind of semi-narcosis in
frogs.[313]

In another investigation, an extract, which also was not chemically defined, showed curare-like activity.[105] An alkaloid has been
isolated by other investigators, but it has not been characterized
chemically, and no pharmacological tests have been carried out.[294]
Seeds of *R. pyramidalis* collected in the province of Oaxaca did
not contain any alkaloids.[169]

The ancient Mexicans may have valued several species of
Rhynchosia as a narcotic.[338] Modern Oaxacan Indians refer to the
toxic seeds of *R. pyramidalis* and *R. longeracemosa* by the same
name—*piule*—that they apply to the seeds of hallucinogenic
morning glories. The black and red *Rhynchosia* beans, pictured
together with mushrooms, have been identified in Aztec paintings,
thus suggesting hallucinogenic use.[349]

> *Rhynchosia longeracemosa* Martens & Galeotti, Bull.
>   Acad. Brux. 10, pt. 2 (1843) 198.
> *Rhynchosia pyramidalis* (*Lam.*) *Urban*, Fedde Rep.
>   15 (1918) 318.

ZYGOPHYLLACEAE

The Zygophyllaceae, belonging to the order Geraniales and
related to the Rutaceae, comprises 25 genera and 250 species of
the tropics and subtropics of both hemispheres. The family,
mostly of woody perennials with many species adapted to xerophytic or halophytic habitats, is divided into five sections.

Peganum Linnaeus

A genus of the Peganoideae—one of the sections of the family
—*Peganum* comprises half a dozen species and occurs from the
Mediterranean eastward to Mongolia and in the southern United
States and Mexico.

The Syrian rue or *P. harmala*, an herb native to dry areas from
the Mediterranean east to northern India, Mongolia and Man-
churia, possesses undoubted hallucinogenic constituents.[347] Its

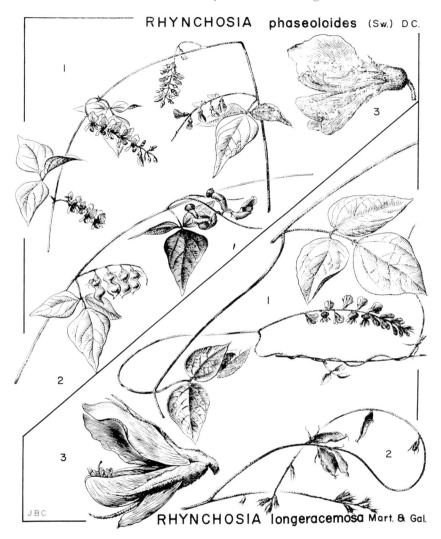

RHYNCHOSIA phaseoloides (Sw.) DC.

RHYNCHOSIA longeracemosa Mart. & Gal.

Figure 72. Drawn by J. B. Clark.

purposeful use as a narcotic to induce visions, however, has not
yet been established with certainty. A critical search of the literature, especially the ancient records, and modern ethnobotanical
field work are needed for a complete understanding of *P.
harmala*.[277]

The seeds of this plant contain the β-carboline alkaloids harmine, harmaline and related bases known to occur in at least eight families of higher plants. (Formulae: see section *Banisteriopsis.*) The fruits of *P. harmala* are the source of a red dye and an oil. This and other species are highly prized in folk medicine as vermifuges, soporifics, alteratives, aphrodisiacs, lactogogues and agents for treating certain eye diseases; the esteem in which *P. harmala* is held amongst peoples of the East is extraordinary and

Figure 73. *Peganum harmala,* south of Beer Sheva, Israel. Photograph by A. Danin.

Phytochemical investigations with *L. inebrians* led to the isolation of a crystalline compound called lagochiline present in the dried plant material up to 3 per cent.[2] The complete chemical structure of lagochiline was elucidated very recently. Lagochiline is a diterpene of the grindelan-type containing 3 primary and 1 secondary hydroxyls and 1 ether oxygen.[64]

Most of the recent pharmacological studies of lagochiline carried out in the Soviet Union deal mainly with its hemostatic effects. There is no information available whether or not lagochiline has any psychotropic activity and whether or not it possesses the physiological activity of the whole plant.[409]

Further ethnobotanical research must be carried out with *L. inebrians* as a basis for more critical and extensive phytochemical and pharmacological studies, especially to ascertain whether or not this inebriating plant is taken for hallucinogenic effects.

Lagochilus inebrians *Bunge,* Mém. Sav. Etr. Petersb. 7 (1847) 438.

SOLANACEAE

Cestrum Linnaeus

The 150 species of *Cestrum* are indigenous to the warmer parts of the Americas.

A species of *Cestrum,* probably *C. laevigatum,* a coastal Brazilian species, is reputedly sold as a substitute for marijuana or maconha (*Cannabis sativa*) in the ports of southern Brazil. Known locally as *dama da noite,* it is said to be employed mainly by seafaring personnel.

The characteristic constituents of species of *Cestrum* are saponins. Gitogenin and digitogenin have been isolated from *C. laevigatum* and *C. parqui.* These substances, however, have no psychotropic activity. There is only one report concerning the isolation of an alkaloid; that is, solasonine, found in *C. parqui.*[359a] Solasonine is a glycosidic steroidal alkaloid, which, on hydrolysis, provides the aglycone solasodine and several sugars, glucose, galactose and rhamnose. Nothing is known about hallucinogenic properties of these types of compounds.

Cestrum laevigatum Schlechtendal, Linnaea 7 (1832) 59.

Iochroma Bentham

A genus of some two dozen species of tropical American shrubs. There are two or three vague reports that several species of *Iochroma* (especially *I fuchsioides*) may occasionally be taken for purposes of inducing hallucinations amongst Indians of the Colombian Andes. Further field work must be carried out, however, before these reports are accepted.[348]

Chemical studies of *Iochroma* have apparently not been carried out.

Iochroma fuchsioides Miers in Hooker Lond. J. Bot. 7 (1848) 345

CAMPANULACEAE

A member of the order Campanulatae and allied to the Goodeniaceae, the Campanulaceae comprises some 60 or 70 genera and over 2000 species of temperate, subtropical and tropical highlands of both hemispheres. Most species are perennial herbs, but a few are trees and shrubs. The family is divided into three sections: Campanuloideae, Cyphioideae, Lobelioideae. Some systematists separate the Lobelioideae and treat it as a separate family, Lobeliaceae.

Lobelia Linnaeus

A cosmopolitan genus, mostly tropical and subtropical and developed especially in America, of 300 species, some of which are valued as ornamentals.

Lobelia tupa, a tall, polymorphic herb distributed in the Andes Mountains, is widely recognized as a toxic species. Known in the southern part of its range as *tupa* or *tabaco del diablo*, the plant is reported to be valued in Chile as a narcotic and medicine. The Chilean peasants employ the juice to relieve toothache, and the Mapuche Indians smoke the leaves as an intoxicant. There is as yet no certainty that the narcotic effects are hallucinogenic, but reports of the methods and purposes of use suggest that they

LOBELIA Tupa L.

Figure 78. Drawn by J. B. Clark.

may be psychoactive. Ethnobotanical field work needs to be carried out where the plant is employed as a fumitory.[231a]

The leaves of tupa contain the piperidine alkaloids lobeline, lobelanidine and norlobelanidine. These constituents are not known to have hallucinogenic effects.[347]

Another campanulaceous plant—*Isotoma longiflora*—is an

additive to the hallucinogenic drink prepared from *Trichocereus* in the Peruvian Andes. It is not known whether or not it contains narcotic principles to justify its addition.

Lobelia tupa Linnaeus Sp. Pl. Ed. 2 (1763) 1318.
Isotoma longiflora Presl Prod. Lobel. (1836) 42.

132. Harner, M. J.: The sound of rushing water. *Nat. Hist.*, 77:28–33, 60–61, 1968.

133. Hartwich, C.: *Die menschlichen Genussmittel*. Chr. Herm. Tauchnitz, Leipzig. 1911.

134. Harvey, D. G. and Robson, W.: The synthesis of r-6-methoxytryptophan and of harmine, with a note on the action of acetaldehyde on tryptophan. *J. Chem. Soc. [Org.]*, *1938*:97–101.

135. Heffter, A.: Ueber zwei Kakteenalkaloide. *Ber. Deutsch. Chem. Ges.*, 27:2975–2979, 1894.

136. ——— Ueber Cacteenalkaloide. *Ber. Deutsch. Chem. Ges.*, 29:216–227, 1896.

137. ——— Ueber Pellote. *Arch. Exp. Pathol. Pharmakol.*, 40:385–429, 1898.

138. Heim, R.: Les champignons divinatoires utilisés dans les rites des indiens Mazatèques, recueillis au cours de leur premier voyage au Mexique, en 1953, par Mme. Valentina Pavlovna Wasson et M. R. Gordon Wasson. *Comptes Rend.*, *242*:965–968, 1956.

139. ——— Les champignons divinatoires recueillis par Mme. Valentina Pavlovna Wasson et M. R. Gordon Wasson au cours de leurs missions de 1954 et 1955 dans les pays Mije, Mazatèque, Zapotèque et Nahua du Mexique Meridional et Central. *Comptes Rend.*, *242*:1389–1395, 1956.

140. ——— Les agarics hallucinogènes du genre *Psilocybe* recueillis au cours de nôtre récente mission dans le Mexique Meridional et Central en compagnie de M. R. Gordon Wasson. *Comptes Rend.*, *244*:659–700, 1957.

141. ——— Notes préliminaires sur les agarics hallucinogènes du Mexique. *Rev. Mycol.*, 22:58–79, 183–198, 1957.

142. ——— *Les Champignons Toxiques et Hallucinogènes*. Paris, N. Boubée & Cie., 1963.

143. ——— Les substances indoliques produites par les champignons toxiques et hallucinogènes. *Bull. Méd. Leg.*, 8:122–141, 1965.

144. ——— *Nouvelles Investigation sur les Champignons Hallucinogènes*. Edit. Mus. Nat. Hist. Nat., Paris, 1967.

145. ——— *Champignons d'Europe*, 2nd ed. Paris, N. Boubée & Cie., 1969.

146. ——— and Hofmann, A.: Phytochimie.—Isolement de la psilocybine à partir du *Stropharia cubensis* Earle et d'autres espèces de champignons hallucinogènes mexicains appartenant au genre *Psilocybe. Comptes Rend.*, 247:557–561, 1958.

147. ——— and Wasson, R. G.: *Les Champignons Hallucinogènes du Mexique*. Edit. Mus. Nat. Hist. Nat., Paris, 1958.

148. ——— and Wasson, R. G.: Une investigation sur les champignons sacrés des Mistèques. *Comptes Rend.*, 254:788–791, 1962.

148a. Heimann, H.: *Die Scopolaminwirkung*, Basel, S. Karger, 1952.

149. Hennings, P.: Eine giftige Kaktee, *Anhalonium Lewinii* n.sp. *Gartenfl.*, 37:410–412, 1888.

149a. Henry, T. A.: *The Alkaloids.* London, J. and A. Churchill, 1949, p. 66.

150. Hernández, F.: *Nova Plantarum, Animalium et Mineralium Mexicanorum Historia. . . .* Rome, B. Deuersini et Z. Masotti, 1651.

150a. Hesse, M.: *Indolalkaloide in Tabellen,* Berlin, Springer-Verlag, 1968, p. 30.

150b. Hesse, O.: *Ueber die Alkaloide der Mandragorawurzel, J. Prakt. Chemie,* 64:274–286, 1901.

151. Hill, R. K.: The *Erythrina* alkaloids. In Manske, R. H. F. (Ed.): *The Alkaloids.* New York, Academic, 1967, vol. 9, pp. 483–515.

152. Hochstein, F. A. and Paradies, A. M.: Alkaloids of *Banisteria caapi* and *Prestonia amazonicum. J. Am. Chem. Soc.,* 79:5735–5736, 1957.

153. Hocking, G. M.: Harmalae Semen. *Quart. J. Crude Drug Res.,* 6:913–915, 1966.

154. Hoffer, A. and Osmond, H.: *The Hallucinogens.* New York, Academic, 1967.

155. Hofmann, A.: Die Geschichte des LSD-25. *Triangel Sandoz-Zeitsch. mediz. Wissen.,* 2:117–124, 1955.

156. Hofmann, A.: Psychotomimetic drugs. Chemical and pharmacological aspects. *Acta Physiol. Pharmacol. Neerl.,* 8:240–258, 1959.

157. ——— Psychotomimetica. Chemische, pharmakologische und medizinische Aspekte. *Svensk Kem. Tidskr.,* 72:723–747, 1960.

158. ——— Chemical, pharmacological and medical aspects of psychotomimetics. *J. Exp. Med. Sci.,* 5:31–51, 1961.

159. ——— Die Wirkstoffe der mexikanischen Zauberdroge "Ololiuqui." *Planta Medica,* 9:354–367, 1961.

160. ——— Psychotomimetic substances. *Indian J. Pharm.,* 25:245–256, 1963.

161. ——— The active principles of the seeds of *Rivea corymbosa* and *Ipomoea violacea. Bot. Mus. Leafl., Harvard U.,* 20:194–212, 1963.

162. ——— Mexikanische Zauberdrogen und ihre Wirkstoffe. *Planta Medica,* 12:341–352, 1964.

163. ——— Alcaloïdes indoliques isolés des plantes hallucinogènes et narcotiques du Mexique. *Colloques Internationaux du Centre National de la Recherche Scientifique, No. 144.* Paris, Editions Centre Nat. Rech. Scient., 1966, pp. 223–241.

164. ——— Psycho-aktive Stoffe aus Pflanzen. *Therapie Woche,* 17:1739–1746. 1967.

165. ——— Psychotomimetic agents. In Burger, A. (Ed.): *Chemical Constitution and Pharmacodynamic Action.* New York, M. Dekker, 1968, vol. 2, ch. 5, pp. 169–235.

166. ——— The discovery of LSD and subsequent investigations on naturally occurring hallucinogens. In Ayd, F. J., Jr. and Blackwell, B. (Eds.): *Discoveries in Biological Psychiatry.* Philadelphia, Lippincott, 1970, ch. 7.

237. Miller, N.G.: The genera of the Cannabaceae in the southeastern United States. *J. Arn. Arb.*, *51*:185–203, 1970.

238. Miras, C. J.: Hashish: its chemistry and pharmacology. Ciba Foundation Study Group, No. 21. London, Churchill, 1965, p. 37.

239. Mitchell, S. W. Remarks on the effects of *Anhalonium lewinii* (the mescal button). *Br. Med. J.*, *1896*, *2*:1625–1629, 1896.

240. Moreau de Tours, J.: *Du Hashisch et de l'Alienation Mentale.* Paris, Masson, 1845.

241. Mors, W. B. and Rizzini, C. T.: *Useful Plants of Brazil.* San Francisco, Holden-Day, 1966.

241a. Mors, W. B. and Ribeiro. O.: Occurrence of scopoletin in the genus *Brunfelsia. J. Org. Chem.*, *22*:978–979, 1957.

242. Morton, C. V.: Notes on yajé, a drug plant of southeastern Colombia. *J. Wash. Acad. Sci.*, *21*:485–488, 1931.

243. Murillo, A.: *Plantes Médicinales du Chili.* Paris, A. Roger et F. Chernoviz, Imprimerie de Lagny, 1889, pp. 152–155.

244. Naranjo, C.: Psychotropic properties of the harmala alkaloids. In Efron, D. H., Holmstedt, B., and Kline, N. S. (Ed.): *Ethnopharmacologic Search for Psychoactive Drugs.* Public Health Serv. Publ. No. 1645, U. S. Govt. Printing Office, Washington, D. C., 1967, pp. 385–391.

245. Naranjo, P.: Etnofarmacología de las plantas psicotrópicas de América, *Terapía*, *24*:5–63, 1969.

246. ———*Ayahuasca: Religión y Medicina.* Quito, Editorial Universitaria, 1970.

247. ——— and de Naranjo, E.: Estudio farmacodinámico de una planta psicotomimética: *Coriaria thymifolia* (shansi). *Arch. de Criminol. Neuro-Psiquiátr. Discipl. Conexas*, *9*:600–616, 1961.

248. O'Connel, F. D. and Lynn, E. V.: The alkaloids of *Banisteriopsis inebrians* Morton. *J. Am. Pharm. Assoc.*, *42*:753–754, 1953.

249. Okuda, S., Kataoka, H., and Tsuda, K.: Studies on lupin alkaloids. III. Absolute configurations of lupin alkaloids. II. *Chem. Pharm. Bull. (Tokyo)*, *13*:491–500, 1965.

250. Ola'h, G.-M.: Etude chimiotaxinomique sur les *Panaeolus.* Recherches sur la présence des corps indoliques psychotropes dans ces champignons. *Comptes Rend.*, *267*:1369–1372, 1968.

251. ——— A taxonomial and physiological study of the genus *Panaeolus* with the Latin descriptions of the new species. *Rev. Mycol.*, *33*:284–290, 1969.

252. ——— *Le genre Panaeolus.* Le Mans, Imprimerie Monnager, 1969.

253. ——— and Heim, R.: Une nouvelle espèce nordaméricaine de *Psilocybe* hallucinogène: *Psilocybe quebecensis. Comptes Rend.*, *264*:1601–1603, 1967.

254. Opler, M. K.: Cross-cultural uses of psychoactive drugs (ethnopharmacology). In Clark, W. G. and del Giudice, J. (Ed.): *Principles of Psychopharmacology.* New York, Academic, 1970.

255. Osmond, H.: Ololiuqui: The ancient Aztec narcotic. Remarks on the effects of *Rivea corymbosa* (ololiuqui). *J. Ment. Sci., 101*:526–537, 1955.

256. Pachter, I. J., Zacharias, D. E., and Ribeiro, O.: Indole alkaloids of *Acer saccharinum* (the silver maple), *Dictyoloma incanescens, Piptadenia colubrina,* and *Mimosa hostilis. J. Org. Chem., 24*: 1285–1287, 1959.

256a. Pachter, I. J. and Hopkinson, A. F.: Note on the alkaloids of *Methysticodendron amesianum. J. Am. Pharm. Assoc., 49*:621–622, 1960.

257. Pardal, R.: Medicina aborigen americana. *Humanior, Sección C, 3:* 1937.

258. Paris, R. and Goutarel, R.: Les Alchornea africains. Présence de yohimbine chez l'*Alchornea floribunda* (Euphorbiacées). *Ann. Pharm. Fr., 16*:15–20, 1958.

259. Parsons, E. C.: *Mitla—Town of the Souls.* Chicago, U. of Chicago, 1936.

260. Pennes, H. H. and Hoch, P. H.: Psychotomimetics, clinical and theoretical considerations: harmine, WIN-2299 and nalline. *Am. J. Psychiatry, 113*:887–892, 1957.

261. Pennington, C. W.: *The Tarahumar of Mexico—Their Environment and Material Culture.* Salt Lake City, U. of Utah, 1963.

262. ——— *The Tepehuan of Chihuahua—Their Material Culture.* Salt Lake City, U. of Utah, 1969.

263. Perkin, W. H. and Robinson, R.: Harmine and harmaline. Part III. *J. Chem. Soc. [Org.], 115*:933–967, 1919.

264. ——— and Robinson, R.: Harmine and harmaline. Part IV. *J. Chem. Soc. [Org.], 115*:967–972, 1919.

265. Perrot, E. and Raymond-Hamet: Le yagé, plante sensorielle des Indiens de la région amazonienne de l'Equateur et de la Colombie. *Comptes Rend., 184*:1266–1268, 1927.

266. ——— and Raymond-Hamet: Yagé, ayahuasca, caapi et leur alcaloïde: telepathine ou yagéine. *Bull. Sci. Pharmacol., 34*:337–347, 417–426, 500–514, 1927.

267. Petrzilka, T., Haefliger, W., and Sikemeier, C.: Synthese von Haschisch-Inhaltsstoffen. *Helv. Chim. Acta, 52*:1102–1134, 1969.

268. Pfeiffer, L.: *Abbildung und Beschreibung blühender Cacteen.* Theodor Fischer, *Cassel. 2:* t. 21, 1848.

269. Philippi, R. A.: *Latua* Ph., ein Neues Genus aus Solanaceen. *Bot. Zeit., 16:No. 33*:241–242, 1858.

269a. Phokas, G. K.: *Contribution to the Definition of the Drastic Components of Mandrake Root.* Thesis, Athens, 1959.

270. Pletscher, A., Besendorf, H., Bächtold, H. P., and Geigy, K. F.: Ueber pharmakologische Beeinflussung des Zentralnervensystems durch kurzwirkende Monoaminoxydasehemmer aus der Gruppe

der Harmala-Alkaloide. *Helv. Physiol. Pharmacol. Acta,* 17:202–214, 1959.

271. Plowman, T., Gyllenhaal, L. O., and Lindgren, J.-E.: *Latua pubiflora:* Magic plant from southern Chile. *Bot. Mus Leafl., Harvard U.,* 23:61–92, 1971.

272. Plugge, P. C. and Rauwerde, A.: Fortgesetzte Untersuchungen über das Vorkommen von Cytisin in verschiedenen Papilionaceae. *Arch. Pharm.,* 234:685–697, 1896.

273. Poisson, J.: Note sur le "natem," boisson toxique péruvienne et ses alcaloïdes. *Ann. Pharm. Fr.,* 23:241–244, 1965.

274. Poisson, M. J.: Présence de mescaline dans une Cactacée péruvienne. *Ann. Pharm. Fr.,* 18:764–765, 1960.

275. Pope, H. G., Jr.: *Tabernanthe iboga*—an African narcotic plant of social importance. *Econ. Bot.,* 23:174–184, 1969.

276. Popelak, A. and Lettenbauer, G.: The mesembrine alkaloids. In Manske, R. H. F. (Ed.): *The Alkaloids.* New York, Academic, 1967, vol. 9, pp. 467–482.

277. Porter, D. M.: The taxonomic and economic uses of *Peganum harmala* (*Zygophyllaceae*) Ms. ined., Botanical Museum, Harvard U., 1962.

278. Prance, G. T.: Notes on the use of plant hallucinogens in Amazonian Brazil. *Econ. Bot.,* 24:62–68, 1970.

279. ――― and Prance, A. E.: Hallucinations in Amazonia. *Gard J.,* 20:102–105, 1970.

280. Prentiss, D. W. and Morgan, F. P.: *Anhalonium lewinii* (mescal buttons). A study of the drug, with especial reference to its physiological action upon man, with report of experiments. *Therap. Gaz.,* 19:577–585, 1895.

281. Raffauf, R. F.: *A Handbook of Alkaloids and Alkaloid-Containing Plants.* New York, Wiley-Interscience, 1970.

282. Ravicz, R.: La Mixteca en el estudio comparativo del hongo alucinante. *An. Inst. Nac. Antrop. Hist.,* 13:73–92, 1960 [1961].

283. Raymond-Hamet and Goutarel, R.: L'*Alchornea floribunda* Mueller-Arg. doit-il à la yohimbine ses effets excitants chez l'homme? *Comptes Rend.,* 261:3223–3224, 1965.

284. Reichel-Dolmatoff, G.: Notes on the cultural extent of the use of yajé (*Banisteriopsis caapi*) among the Indians of the Vaupés, Colombia. *Econ. Bot.,* 24:32–33, 1970.

285. Reinberg, P.: Contribution à l'étude des boissons toxiques des Indiens du nord-ouest de l'Amazone, l'ayahuasca, le yajé, le huanto. *J. Soc. Am. Paris, n.s.,* 13:25–54, 197–216, 1921.

286. Reiner, R. and Eugster, C. H.: Zur Kenntnis des Muscazons. *Helv. Chim. Acta,* 50:128–136, 1967.

287. Reko, B. P.: De los nombres botánicos aztecos. *El Méx. Ant., 1,* No. 5:136, 152, 1919.

288. ——— Alcaloides y glucosidos en plantas mexicanas. *Mem. Soc. Alzate*, 49:412, 1929.

289. ——— Das mexikanische Rauschgift Ololiuqui. *El Méx. Ant.*, 3, No. 3–4, 1–7, 1934.

290. Reko, V. A.: *Magische Gifte—Rausch- und Betäubungsmittel der Neuen Welt*. Stuttgart, Ferdinand Enke Verlag, 1936.

291. Reti, L.: Cactus alkaloids and some related compounds. In *Progress in the Chemistry of Organic Natural Products*. Vienna, Springer-Verlag, 1950, vol. 6, pp. 242–289.

292. ——— and Castrillon, J. A.: Cactus alkaloids. I. *Trichocereus terscheckii* (Parmentier) Britton and Rose. *J. Am. Chem. Soc.*, 73:1767–1769, 1951.

293. Ríos, O.: Aspectos preliminares al estudio farmacopsiquiátrico del ayahuasca y su principio activo. *An. Fac. Med. U. Nac. Mayor S. Marcos*, 45:22–66, 1962.

294. Ristic, S. and Thomas, A.: Zur Kenntnis von *Rhynchosia pyramidalis* (Pega-Palo). *Arch. Pharmaz.*, 295:510, 1962.

295. Ritchie, E. and Taylor, W. C.: The galbulimima alkaloids. In Manske, R. H. F. (Ed.): *The Alkaloids*. New York, Academic, 1967, vol. 9, pp. 529–543.

295a. Rivier, L. and Lindgren, J.-E.: Ayahuasca-South American hallucinogenic drink: ethnobotanical and chemical investigations. *Econ. Bot.*, in press.

296. Robbers, J. E., Tyler, V. E., and Ola'h, G. M.: Additional evidence supporting the occurrence of psilocybin in *Panaeolus foenisecii*. *Lloydia*, 32:399–400, 1969.

297. Robichaud, R. C., Malone, M. H., and Kosersky, D. S.: Pharmacodynamics of cryogenine, an alkaloid isolated from *Heimia salicifolia* Link and Otto. Part II. *Arch. Int. Pharmacodyn. Ther.*, 157:43–52, 1965.

298. ———, Malone, M. H., and Schwarting, A. E.: Pharmacodynamics of cryogenine, an alkaloid isolated from *Heimia salicifolia* Link and Otto. Part I. *Arch. Int. Pharmacodyn. Ther.*, 150:220–232, 1964.

299. Rouhier, A.: *La Plante qui Fait les Yeux Emerveillés-le Peyotl*. Paris, Gaston Doin et Cie., 1927.

300. Rusby, H. H.: The aboriginal uses of caapi. *J. Am. Pharm. Assoc.*, 12:1123, 1923.

301. Safford, W. E.: An Aztec narcotic. *J. Hered.*, 6:291–311, 1915.

302. ——— Identity of cohoba, the narcotic snuff of ancient Haiti. *J. Wash. Acad. Sci.*, 6:548–562, 1916.

303. ——— Narcotic plants and stimulants of the ancient Americans. *Ann. Rep. Smithson. Inst.*, 1916:387–424, 1917.

304. ——— Daturas of the Old World and New. *Ann. Rep. Smithson. Inst.*, 1920:537–567, 1920.

305. ———— *Datura*—an inviting genus for the study of heredity. *J. Hered.*, 12:178–190, 1921.

306. ———— Synopsis of the genus *Datura*. *J. Wash. Acad. Sci.*, 11:173–189, 1921.

307. ———— Daturas of the Old World and New: an account of their narcotic properties and their use in oracular and initiatory ceremonies. *Ann. Rep. Smithson. Inst.*, 1920:537–567, 1922.

308. Sahagún, B.: *Historia General de las Cosas de la Nueva España.* Mexico, D. F., Editorial Pedro Robledo, 1938, vol. 3.

309. Sai-Halás-z, A., Brunecker, G., and Szàra, S.: Dimethyltryptamin: ein neues Psychoticum. *Psychiatr. Neurol.*, (*Basel*), 135:285–301, 1958.

310. Santavý, F.: *Acta Univ. Palackianae Olomuc, Fac. Med.*, 35 (1964) 5.

311. Santesson, C. G.: Notiz über piule, eine mexikanische Rauschdroge. *Ethnol. Stud.* (*Gothenburg*), 4:1–11, 1937.

312. ———— Piule, eine mexikanische Rauschdroge. *Arch. Pharm. 1937*: 532–537, 1937.

313. ———— Noch eine mexikanische "Piule"-Droge. Semina Rynchosiae phaseoloidis DC. *Ethnolog. Studier*, 6:179–183, 1938.

314. Satina, S. and Avery, A. G.: A review of the taxonomic history of *Datura*. In Avery, A. G., Satina, S., and Rietsema, J. (Eds.): *Blakeslee: The Genus Datura.* New York, Ronald, 1959, pp. 16–47.

315. Schmiedeberg, O. and Koppe, R.: *Das Muscarin, das giftige Alkaloid des Fliegenpilzes*, Leipzig, Verlag Vogel, R. C., 1869.

316. Schneider, J. A. and Sigg, E. B.: Neuropharmacological studies on ibogaine. *Ann. N. Y. Acad. Sci.*, 66:765, 1957.

317. ———— and Sigg, E. B.: Pharmacologic analysis of tranquilizing and central stimulating effects. In Pennes, H. H. (Ed.): *Psychopharmacology.* New York, Hoeber, 1958, ch. 4, pp. 75–98.

318. Schultes, R. E.: Peyote and plants used in the peyote ceremony. *Bot. Mus. Leafl., Harvard U.*, 4:129–152, 1937.

319. ———— Peyote (*Lophophora williamsii*) and plants confused with it. *Bot. Mus. Leafl., Harvard U.*, 5:61–88, 1937.

320. ———— *Peyote* (*Lophophora williamsii* (*Lemaire*) *Coulter*) *and Its Uses*. Thesis ined., Harvard U., 1937.

321. ———— Plantae Mexicanae II. The identification of teonanacatl, a narcotic Basidiomycete of the Aztecs. *Bot. Mus. Leafl., Harvard U.*, 7:37–54, 1939.

322. ———— The appeal of peyote (*Lophophora williamsii*) as a medicine. *Am. Anthrop.*, 40:698–715, 1938.

323. ———— Teonanacatl, the narcotic mushroom of the Aztecs. *Am. Anthrop.*, 42:429–443, 1940.

324. ———— The aboriginal therapeutic uses of *Lophophora Williamsii*. *Cact. Succ. J.*, 12:177–181, 1940.

325. ———— A contribution to our knowledge of Rivea corymbosa, the narcotic ololiuqui of the Aztecs. Cambridge, Mass., Harvard Botanical Museum, 1941.

326. ————Plantae Austro-Americanae IX. *Bot. Mus Leafl., Harvard U., 16:*202–205, 1954.

327. ———— A new narcotic snuff from the northwest Amazon. *Bot. Mus. Leafl., Harvard U., 16:*241–260, 1954.

328. ———— A new narcotic genus from the Amazon slope of the Colombian Andes. *Bot. Mus. Leafl., Harvard U., 17:*1–11, 1955.

329. Schultes, R. E.: The identity of the malpighiaceous narcotics of South America. *Bot. Mus. Leafl., Harvard U. 18:*1–56, 1957.

330. ———— Botany attacks the hallucinogens. *Texas J. Pharm., 2:*168–185, 1961.

331. ———— Native narcotics of the New World. *Texas J. Pharm., 2:*141–167, 1961.

332. ———— Tapping our heritage of ethnobotanical lore. *Econ. Bot., 14:*257–262, 1960; *Chemurgic Dig., 20:*10–12. 1961.

333. ———— The role of the ethnobotanist in the search for new medicine plants. *Lloydia, 25:*257–266, 1962.

334. ———— Botanical sources of the New World narcotics. *Psyched. Rev., 1:*145–166, 1963.

335. ———— Hallucinogenic plants of the New World. *Harvard Rev., 1:* 18–32, 1963.

336. ———— The widening panorama in medical botany. *Rhodora, 65:*97–120, 1963.

337. ———— The correct names for two Mexican narcotics. *Taxon, 13:*65–66, 1964.

338. ———— Ein Halbes Jahrhundert Ethnobotanik amerikanischer Halluzinogene. *Planta Medica, 13:*126–157, 1965.

339. ———— The search for new natural hallucinogens. *Lloydia, 29:*293–308, 1966.

340. ———— The place of ethnobotany in the ethnopharmacological search for psychotomimetic drugs. In Efron, D., Holmstedt, B., and Kline, N. S. (Eds.): *Ethnopharmacological Search for Psychoactive Drugs.* Public Health Serv. Publ. No. 1645, U. S. Govt. Printing Office, Washington, D. C., 1967, pp. 33–57.

341. ———— The botanical origins of South American snuffs. In Efron, D., Holmstedt, B., and Kline, N. S. (Eds.): *Ethnopharmacologic Search for Psychoactive Drugs.* Public Health Serv. Publ. No. 1645, U. S. Govt. Printing Office, Washington, D. C., 1967, pp. 291–306.

342. ———— The plant kingdom and modern medicine. *Herbalist, No. 34:* 18–26, 1968.

343. ———— Some impacts of Spruce's explorations on modern phytochemical research. *Rhodora, 70:*313–339, 1968.

344. ———— De plantis toxicariis e Mundo Novo tropicale commentationes

IV. *Virola* as an orally administered hallucinogen. *Bot. Mus. Leafl., Harvard U.,* 22:133–164, 1969.

345. ——— Hallucinogens of plant origin. *Science, 163*:245–254, 1969.

346. ——— The unfolding panorama of New World hallucinogens. In Gunckel, J. E. (Ed.): *Current Topics in Plant Science.* New York, Academic, pp. 336–354, 1969.

347. ——— The botanical and chemical distribution of hallucinogens. *Ann. Rev. Pl. Physiol.,* 21:571–594, 1970.

348. ——— The New World Indians and their hallucinogenic plants. *Bull. Morris Arb.,* 21:3–14, 1970.

349. ——— The plant kingdom and hallucinogens. *Bull. Narcotics, 21,* No. 3:3–16; No. 4:15–27, 1969; 22, No. 1:25–53, 1970.

349a. ——— The utilization of hallucinogens in primitive societies—use, misuse or abuse? In Keup, W. (Ed.): *Drug Abuse. Current Concepts and Research.* Springfield, Ill., Charles C Thomas, Publishers, 17–26, 1972.

350. ——— and Holmstedt, B.: De plantis toxicariis e Mundo Novo tropicale commentationes II. The vegetal ingredients of the myristicaceous snuffs of the northwest Amazon. *Rhodora, 70*:113–160, 1968.

351. ———, Holmstedt, B., and Lindgren, J.-E.: De plantis toxicariis e Mundo Novo tropicale commentationes. III. Phytochemical examination of Spruce's original collection of *Banisteriopsis caapi. Bot. Mus. Leafl., Harvard U.,* 22:121–132, 1969.

352. ——— and Raffauf, R. F.: *Prestonia;* an Amazon narcotic or not? *Bot. Mus. Leafl., Harvard U.,* 19:109–122, 1960.

353. Schultz, O. E. and Haffner, G.: Zur Kenntnis eines sedativen und antibakteriellen Wirkstoffes aus dem deutschen Faserhanf (*Cannabis sativa*). *Z. Naturforsch., 14b*:98–100, 1959.

354. ——— and Haffner, G.: Zur Kenntnis eines sedativen Wirkstoffes aus dem deutschen Faserhanf (*Cannabis sativa*). *Arch. Pharmaz., 291*:391–403, 1958.

355. Schulz, B.: *Lagochilus inebrians* Bge., eine interessante neue Arzneipflanze. *Dtsch. Apotheker-Zeit.,* 99:1111, 1959.

356. Seitz, G.: Epená, the intoxicating snuff powder of the Waiká Indians and the Tucano medicine man, Agostino. In Efron, D., Holmstedt, B., and Kline, N. S. (Eds.): *Ethnopharmacologic Search for Psychoactive Drugs.* Public Health Serv. Publ. No. 1645, U. S. Govt. Printing Office, Washington, D. C., 1967, pp. 315–338.

357. Shinners, L.: Correct nomenclature of two Mexican narcotic plants. *Taxon, 14*:103–105, 1965.

358. Shonle, R.: Peyote—giver of visions. *Am. Anthrop.,* 27:53–75, 1925.

359. Shulgin, A. T., Sargent, T., and Naranjo, C.: The chemistry and psychopharmacology of nutmeg and several related phenylisopropylamines. In Efron, D. H., Holmstedt, B., and Kline, N. S.

(Eds.): *Ethnopharmacologic Search for Psychoactive Drugs.* Public Health Serv. Publ. No. 1645, U. S. Government Printing Office, Washington, D. C., 1967, pp. 202–214.

359a. Silva, M., Mancinelli, P., and Cheul, M.: Chemical Study of *Cestrum parqui. J. Pharm. Sci., 51*:289, 1962.

360. Silverwood-Cope, P.: Personal communication.

361. Singer, R.: Mycological investigations on teonanacatl, the Mexican hallucinogenic mushroom. Part I. The history of teonanacatl, field work and culture work. *Mycologia, 50*:239–261, 1958.

362. ———— Pilze, die Zerebralmyzetismen Verursachen. *Schweiz-Z. Pilzkunde, 36*:81–89, 1958.

363. Singer, R.: *The Agaricales in Modern Taxonomy,* 2nd ed. Weinheim, J. Cramer, 1962.

364. ———— and Smith, A. H.: Mycological investigations on teonanacatl, the Mexican hallucinogenic mushroom. Pt. II. A taxonomic monograph of *Psilocybe,* section *Caerulescentes. Mycologia, 50*:262- 303, 1958.

365. Sirakawa, K., Aki, O., Tsushima, S., and Konishi, K.: Synthesis of ibotenic acid and 3-deoxyibotenic acid. *Chem. Pharm. Bull. (Tokyo), 14*:89–91, 1966.

366. Slotkin, J. S.: *The Peyote Religion.* Glencoe, Ill., Free Press, 1956.

367. Slotta, K. H. and Heller, H.: Ueber β-Phenyläthylamin. I. Mitteil.: Mezcalin und mezcalinähnliche Substanzen. *Ber., 63*:3029–3044, 1930.

368. ———— and Szyszka, G.: Ueber β-Phenyläthylamine. III. Mitt.: Neue Darstellung von Mescalin. *J. Prakt. Chem., 137*:339–350, 1933.

369. Smith, S. and Timmis, G. M.: The alkaloids of ergot. Part III. Ergine, a new base obtained by the degradation of ergotoxine and ergotinine. *J. Chem. Soc.* [*Org.*], *1932*:763–766, 1932.

370. ———— and Timmis, G. M.: The alkaloids of ergot. Part VII. Isoergine and isolysergic acid. *J. Chem. Soc.* [*Org.*], *1936*:1440–1444, 1936.

371. Söderblom, N.: *Rus och Religion.* Bokfenix, Uppsala, 1968.

372. Solms, H.: Chemische Struktur und Psychose bei Lysergsäure-Derivaten. *Praxis, 45*:746–749, 1956.

373. ———— Relationships between chemical structure and psychoses with the use of psychotoxic substances. *J. Clin. Exp. Psychopath. Quart. Rev. Psychiat. Neurol., 17*:429–433, 1956.

374. Späth, E.: Ueber die Anhalonium-Alkaloide. *Monatsh. Chemie (Wien), 40*:129–154, 1919.

375. ———— Ueber das Carnegin. *Ber. Dtsch. Chem. Ges., 62*:1021–1024, 1929.

376. ———— and Bruck, J.: Ueber ein neues Alkaloid aus den Mezcalbuttons (XVIII. Mitteilung über Kakteen-Alkaloide). *Ber. Dtsch. Chem. Ges., 70*:2446–2450, 1937.

377. ———— and Bruck, J.: N-Acetyl-mezcalin als Inhaltsstoff der Mezcal-buttons (XIX. Mitteilung über Kakteen-Alkaloide). *Ber. Dtsch. Chem. Ges., 71*:1275–1276, 1938.

378. ———— and Lederer, E.: Synthese der Harmala-Alkaloide: Harmalin, Harmin and Harman. *Ber., 63*:120–125, 1930.

379. ———— and Lederer, E.: Synthesen von 4-Carbolinen. *Ber., 63*:2102–2111, 1930.

380. Speeter, M. E. and Anthony, W. C.: The action of oxalyl chloride on indoles: a new approach to tryptamines. *J. Am. Chem. Soc., 76*: 6208–6210, 1954.

381. Spenser, I. D.: A synthesis of harmaline. *Can. J. Chem., 37*:1851–1858, 1959.

382. Spruce, R.: In Wallace, A. R. (Ed.): *Notes of a Botanist on the Amazon and Andes.* London, Macmillan, 1908. (Reprinted ed.) New York, Johnson Reprint, 1970.

382a. Staub, H.: Ueber die chemischen Bestandteile der Mandragora-wurzel. 2. Die Alkaloide. *Helv. Chim. Acta, 45*:2297–2305, 1962.

383. Stevenson, M. C.: Ethnobotany of the Zuni Indians. *30th Ann. Rep. Bur. Am. Ethnol.*, pp. 31–102, 1908–09.

384. ———— The Zuni Indians: their mythology, esoteric fraternities and ceremonies. *23rd Ann. Rep. Bur. Am. Ethnol.*, pp. 1–634, 1901–02.

385. Stoll, A. and Hofmann, A.: Partialsynthese von Alkaloiden von Typus des Ergobasins. *Helv. Chim. Acta, 26*:944–965, 1943.

386. ————, Hofmann, A., and Schlientz, W.: Die stereoisomeren Lyser-gole und Dihydro-lysergole. *Helv. Chim. Acta, 32*:1947–1956, 1949.

386a. ———— and Jucker, E.: Neuere Untersuchungen von natürlichen und synthetischen Tropan-Derivaten. *Angewandte Chemie* [*Weinheim*], *66*:376–386, 1954.

387. ————, Troxler, F., Peyer, J., and Hofmann, A.: Eine neue Synthese von Bufotenin und verwandten Oxy-tryptaminen. *Helv. Chim. Acta, 38*:1452–1472, 1955.

388. Stowe, B.: Occurrence and metabolism of simple indoles in plants. *Fortschr. Chem. Org. Naturst., 17*:248–297, 1959.

389. Stromberg, V. L.: The isolation of bufotenine from *Piptadenia peregrina. J. Am. Chem. Soc., 76*:1707, 1954.

390. St. Szára: Dimethyltryptamin: Its metabolism in man; the relation of its psychotic effect to the serotonin metabolism. *Experientia, 12*:441–442, 1956.

391. ———— Hallucinogenic effects and metabolism of tryptamine derivatives in man. *Fed. Proc., 20*:885–888, 1961.

392. Taber, W. A. and Heacock, R. A.: Location of ergot alkaloid and fungi in the seed of *Rivea corymbosa* (L.) Hall.f., "Ololiuqui." *Can. J. Microbiol., 8*:137–143, 1962.

393. ——, Heacock, R. A., and Mahon, M. E.: Ergot-type alkaloids in vegetative tissue of *Rivea corymbosa* (L.) Hall.f. *Phytochem.*, 2:99–101, 1963.

394. ——, Vining, L. C., and Heacock, R. A.: Clavine and lysergic acid alkaloids in varieties of Morning Glory. *Phytochem.*, 2:65–70. 1963.

395. Takemoto, T., Nakajima, T., and Yokobe, T.: Structure of ibotenic acid. *J. Pharm. Soc. Jap.*, 84:1232–1233, 1964.

396. ——, Yokobe, T., and Nakajima, T.: Studies on the constituents of indigenous fungi. II. Isolation of the flycidal constituent from *Amanita strobiliformis*. *J. Pharm. Soc. Jap.*, 84:1186–1188, 1964.

397. Taylor, N.: *Flight from Reality.* New York, Duell, Sloan and Pearce, 1949.

398. Taylor, W. I.: *Indole Alkaloids.* Oxford, Pergamon, 1966.

399. Theobald, W., Büch, O., Kunz, H. A., Krupp, P., Stenger, E. G., and Heimann, H.: Pharmakologische und experimentalpsychologische Untersuchungen mit 2 Inhaltsstoffen des Fliegenpilzes (*Amanita muscaria*). *Arzneim.-Forsch.*, 18:311–315, 1968.

399a. Todd, James S.: Thin-layer chromatography analysis of Mexican population of Lophophora (Cactaceae). *Lloydia*, 32:395–398, 1969.

400. Troike, R. C.: The origin of Plains mescalism. *Am. Anthrop.*, 64:946–963, 1962.

401. Troxler, F., Seemann, F., and Hofmann, A.: Abwandlungsprodukte von Psilocybin und Psilocin. *Helv. Chim. Acta*, 42:2073–2103, 1959.

402. Truitt, E. B., Jr., The pharmacology of myristicin and nutmeg. In Efron, D. H., Holmstedt, B., and Kline, N. S. (Eds.): *Ethnopharmacologic Search for Psychoactive Drugs.* Public Health Serv. Publ. No. 1645, U. S. Govt. Printing Office, Washington, D. C., 1967, pp. 215–222.

403. ——, Callaway, C., Braude, M. C., and Krantz, J. C., Jr.: The pharmacology of myristicin. A contribution to the psychopharmacology of nutmeg. *J. Neuropsych.*, 2:205–210, 1961.

404. Tsao, U. T.: A new synthesis of mescaline. *J. Am. Chem. Soc.*, 73:5495–5496, 1951.

405. Turner, W. J. and Heyman, J. J.: The presence of mescaline in *Opuntia cylindrica. J. Org. Chem.*, 25:2250, 1960.

406. —— and Merlis, S.: Effect of some indolealkylamines on man. *Arch. Neurol. Psychiat.*, 81:121–129, 1959.

407. ——, Merlis, S., and Carl, A.: Concerning theories of indoles and schizophrenigenesis. *Am. J. Psychiatry*, 112:466–467, 1955.

408. Tyler, V. E.: Indole derivatives in certain North American mushrooms. *Lloydia*, 24:71–74, 1961.

409. —————— The physiological properties and chemical constituents of some habit-forming plants. *Lloydia*, 29:275–291, 1966.

410. Uhle, M.: A snuffing tube from Tiahuanaco. *Bull. Free Mus. Sci. Art*, U. of Penn., *1, No. 4*:158–177, 1898.

411. Urbina, M.: *Catálogo de Plantas Mexicanas (Fanerógamas)*. Mexico, D. F., 1897.

412. —————— El peyote y el ololiuqui. *An. Mus. Nac. México*, 7:25–38, 1903; *La Naturaleza, 1, No. 4*:1912.

413. Uscátegui-M, N.: The present distribution of narcotics and stimulants amongst the Indian tribes of Colombia. *Bot. Mus. Leafl., Harvard U., 18*:273–304, 1959.

414. Villavicencio, M.: *Geografía de la República del Ecuador*, New York, R. Craigshead, 1858, p. 371.

415. Vincent, D. and Sero, J.: Action inhibitrice de *Tabernanthe iboga* sur la cholinestérase du sérum. *C. R. Soc. Biol., 136*:612–614, 1942.

416. Vitali, T. and Mossini, F.: Sulla preparazione di alcune triptamine N'-disostituite. *Boll. Sci. Fac. Chim. Ind. Univ. Bologna, 17*:84–87, 1959.

417. Vrkoc, J., Herout, V., and Sorm, F.: On terpenes. CXXXIII. Structure of acorenone, a sesquiterpenic ketone from sweet-flag oil (*Acorus calamus* L.) *Coll. Czechoslov. Chem. Commun., 26*:3183–3185, 1961.

418. Wagner, H.: *Rauschgift-Drogen*. Berlin, Springer Verlag, 1969.

419. Warburg, O.: *Die Muscatnuss*. Leipzig, Verlag von Wilhelm Engelmann, 1897.

420. Waser, P. G.: The pharmacology of *Amanita muscaria*. In Efron, D. H., Holmstedt, B., and Kline, N. S. (Eds.): *Ethnopharmacologic Search for Psychoactive Drugs*. Public Health Service Publ. No. 1645, U. S. Govt. Printing Office, Washington, D. C., 1967, pp. 419–439.

421. Wassén, S. H.: Some general viewpoints in the study of native drugs, especially from the West Indies and South America. *Ethnos*, 1–2:97–120, 1964.

422. —————— The use of some specific kinds of South American Indian snuff and related paraphernalia. *Etnolog. Studier, 28*:1–116, 1965.

423. —————— Anthropological survey of the use of South American snuffs. In Efron, D. H., Holmstedt, B., and Kline, N. S. (Eds.): *Ethnopharmacologic Search for Psychoactive Drugs*. Public Health Service Publ. No. 1645, U. S. Govt. Printing Office, Washington, D. C., 1967, pp. 233–289.

424. —————— Om bruket av hallucinogena snuser av sydamerikanskt ursprung. *Sydsvenska Medicin-Historiska Sälsk. Årsschrift*, 70–98, 1969.

425. ———— and Holmstedt, B.: The use of paricá, an ethnological and pharmacological review. *Ethnos, 1:5–45*, 1963.

426. Wasson, R. G.: The divine mushroom: primitive religion and hallucinatory agents. *Proc. Am. Phil. Soc., 102:221–223*, 1958.

427. ———— The hallucinogenic mushrooms of Mexico and psilocybin: a bibliography. *Bot. Mus. Leafl., Harvard U., 20:25–73*, 1962.

428. ———— A new Mexican psychotropic drug from the Mint Family. *Bot. Mus. Leafl., Harvard U., 20:77–84*, 1962.

429. ———— Notes on the present status of ololiuhqui and other hallucinogens of Mexico. *Bot. Mus. Leafl., Harvard U., 20:161–193*, 1963.

430. ———— *Soma, Divine Mushroom of Immortality.* New York, Harcourt, 1967.

431. ———— The fly agaric and man. In Efron, D., Holmstedt, B., and Kline, N. S. (Eds.): *Ethnopharmacologic Search for Psychoactive Drugs.* Public Health Service Publ. No. 1645, U. S. Govt. Printing Office, Washington, D. C., 1967, pp. 405–414.

432. ———— Soma of the Aryans: an ancient hallucinogen. *Bull. Narcotics, 22, No. 3:25–30*, 1970.

433. Wasson, V. P. and Wasson, R. G.: *Mushrooms, Russia and History.* New York, Pantheon, 1957.

434. Watt, J. M. and Breyer-Brandwijk, M. G.: *The Medicinal and Poisonous Plants of Southern and Eastern Africa,* 2nd ed. Edinburgh, Livingstone, 1962, p. 759.

435. Weidmann, H., Taeschler, M., and Konzett, H.: Zur Pharmakologie von Psilocybin, einem Wirkstoff aus *Psilocybe mexicana* Heim. *Experientia, 14:378–379*, 1958.

436. ———— and Cerletti, A.: Zur pharmakodynamischen Differenzierung der 4-Oxyindolderivate Psilocybin und Psilocin im Vergleich mit 5-Oxyindolkörpern (Serotonin, Bufotenin). *Helv. Physiol. Acta, 17:C 46–C 48*, 1959.

437. Weil, A. T.: Cannabis. *Sci. J., 5A, No. 3:36–42*, 1969.

438. ———— Nutmeg as a narcotic. *Econ. Bot., 19:194–217*, 1965.

439. ———— Nutmeg as a psychoactive drug. In Efron, D., Holmstedt, B., and Kline, N. S. (Eds.): *Ethnopharmacologic Search for Psychoactive Drugs.* Public Health Service Publ. No. 1645, U. S. Govt. Printing Office, Washington, D. C., 1967, p. 202–214.

440. ———— Nutmeg and other psychoactive groceries. In Gunckel, J.E. (Ed.): *Current Topics in Plant Science.* New York, Academic, 1969, pp. 355–366.

441. ————, Zinberg, N. E., and Nelsen, J. M.: Clinical and psychological effects of marihuana in man. *Science, 162:1234–1242*, 1968.

442. White, O. E.: Botanical explorations in Bolivia. *Brooklyn Bot. Gard. Rec., 11, No. 3:93–105*, 1922.

443. Wieland, T.: *Science, 159:946–952.*

444. Willaman, J. J. and Li, H.-L.: Alkaloid-bearing plants and their contained alkaloids, 1957–1968. *Lloydia, Supplement 33, No. 3A,* 1970.

445. Wolbach, A. B., Isbell, H., and Miner, E. J.: Cross tolerance between mescaline and LSD-25 with a comparison of the mescaline and LSD reactions. *Psychopharmac., 3:*1–14, 1962.

446. ———— Miner, E. J., and Isbell, H.: Comparison of psilocin with psilocybin, mescaline and LSD-25. *Psychopharmac., 3:*219–223, 1962.

447. Wolfes, O. and Rumpf, K.: Ueber die Gewinnung von Harmin aus einer südamerikanischen Liane. *Arch. Pharm., 266:*188–189, 1928.

448. Wurdack, J. J.: Indian narcotics in southern Venezuela. *Gard. J.,* 8:116–118, 1958.

449. Ximénez, F.: *Quatro libros de la naturaleza. . . .* Mexico, D. F., 1615, Lib. 2, ch. 14.

450. Yui, T. and Takeo, Y.: Neuropharmacological studies on a new series of ergot alkaloids. *Jap. J. Pharmacol., 7:*157–161, 1958.

451. Zacharias, D. E., Jeffrey, G. A., Douglas, B., Weisbach, J. A., Kirkpatrick, J. L., Ferris, J. P., Boyce, C. B., and Briner, R. C.: The structure of O-methyllythrine hydrobromide. *Experientia, 21:*247–248, 1965.

452. Zachowski, J.: Zur Pharmakologie des Cytisins. *Arch. Exp. Path. Pharmakol., 189:*327–344, 1938.

453. Zerries, O.: Medizinmannwesen und Geisterglaube der Waiká-Indianer des Oberen Orinoko. *Ethnologica, N. F., 2:*485–507, 1960.

INDEX